Mining and Communities:

Understanding the Context of Engineering Practice

Synthesis Lectures on Engineers, Technology, and Society

Editor

Caroline Baillie, *University of Western Australia*

The mission of this lecture series is to foster an understanding for engineers and scientists on the inclusive nature of their profession. The creation and proliferation of technologies needs to be inclusive as it has effects on all of humankind, regardless of national boundaries, socio-economic status, gender, race and ethnicity, or creed. The lectures will combine expertise in sociology, political economics, philosophy of science, history, engineering, engineering education, participatory research, development studies, sustainability, psychotherapy, policy studies, and epistemology. The lectures will be relevant to all engineers practicing in all parts of the world. Although written for practicing engineers and human resource trainers, it is expected that engineering, science and social science faculty in universities will find these publications an invaluable resource for students in the classroom and for further research. The goal of the series is to provide a platform for the publication of important and sometimes controversial lectures which will encourage discussion, reflection and further understanding.

The series editor will invite authors and encourage experts to recommend authors to write on a wide array of topics, focusing on the cause and effect relationships between engineers and technology, technologies and society and of society on technology and engineers. Topics will include, but are not limited to the following general areas; History of Engineering, Politics and the Engineer, Economics , Social Issues and Ethics, Women in Engineering, Creativity and Innovation, Knowledge Networks, Styles of Organization, Environmental Issues, Appropriate Technology.

Mining and Communities: Understanding the Context of Engineering Practice
Rita Armstrong, Caroline Baillie, and Wendy Cumming-Potvin
February 2014

Engineering and War: Militarism, Ethics, Institutions, Alternatives
Ethan Blue, Michael Levine, Dean Nieusma
December 2013

Engineers Engaging Community: Water and Energy

iv

Engineering and Society: Working Towards Social Justice, Part III: Windows on Society
Caroline Baillie, George Catalano
2009

Engineering: Women and Leadership
Corri Zoli, Shobha Bhatia, Valerie Davidson, Kelly Rusch
2008

Bridging the Gap Between Engineering and the Global World: A Case Study of the Coconut (Coir) Fiber Industry in Kerala, India
Shobha K. Bhatia, Jennifer L. Smith
2008

Engineering and Social Justice
Donna Riley
2008

Engineering, Poverty, and the Earth
George D. Catalano
2007

Engineers within a Local and Global Society
Caroline Baillie
2006

Globalization, Engineering, and Creativity
John Reader
2006

Engineering Ethics: Peace, Justice, and the Earth
George D. Catalano
2006

Mining Communities: Understanding the Context of Engineering Practice
Rita Armstrong, Caroline Baillie, and Wendy Cumming-Potvin

ISBN: 978-3-031-00986-0 print
ISBN: 978-3-031-02114-5 ebook

DOI 10.1007/978-3-031-02114-5

 A Publication in the Springer Nature series
SYNTHESIS LECTURES ON ADVANCES IN AUTOMOTIVE TECHNOLOGY # 21
Series Editor: Caroline Baillie, University of Western Australia

Series ISSN 1933-3633 Print 1933-3641 Electronic

Mining and Communities:

Understanding the Context of Engineering Practice

Rita Armstrong
University Western Australia
Caroline Baillie
University Western Australia
Wendy Cumming-Potvin
Murdoch University

SYNTHESIS LECTURES ON ENGINEERS, TECHNOLOGY, AND SOCIETY # 21

ABSTRACT

Mining has been entangled with the development of communities in all continents since the beginning of large-scale resource extraction. It has brought great wealth and prosperity, as well as great misery and environmental destruction. Today, there is a greater awareness of the urgent need for engineers to meet the challenge of extracting declining mineral resources more efficiently, with positive and equitable social impact and minimal environmental impact. Many engineering disciplines—from software to civil engineering—play a role in the life of a mine, from its inception and planning to its operation and final closure. The companies that employ these engineers are expected to uphold human rights, address community needs, and be socially responsible. While many believe it is possible for mines to make a profit and achieve these goals simultaneously, others believe that these are contradictory aims. This book narrates the social experience of mining in two very different settings—Papua New Guinea and Western Australia—to illustrate how political, economic, and cultural contexts can complicate the simple idea of "community engagement."

KEYWORDS

mining and community engagement, mining and social impact, mining and development, Ok Tedi, Boddington, Pilbara

Contents

Acknowledgments

We would like to extend our gratitude to all the people who made themselves available to be interviewed for our research and particular thanks to Eric Feinblatt for invaluable assistance in recording and filming all interviews. The research for this book was supported by an Australian Learning and Teaching Council Grant CG10-1519 "Engineering Education for Social and Environmental Justice."

Preface

INTRODUCTION

Mining, in all its forms, has attracted both praise and criticism. It has brought wealth and suffering to societies all round the world, yet different perceptions of mining are not necessarily aligned in conventional ways. Opposition to mining may come from left-wing anti-mining lobby groups such as Mines Against Communities (MAC 2013) to otherwise conservative local rural groups who wish to protect their farms, their environment, and their community life from the incursions of, for example, coal seam gas extraction. Some governments (such as Norway) have strong regulations on the mining industry and explicitly set aside revenue for public expenditure while the desire of the Australian government to increase revenue in the form of a mining tax has generated a strong backlash from the powerful mining lobby. Governments often welcome the revenue from mining but there is a vast difference between a government's desire to regulate the actions of mining companies and their ability to do so.

Mining can be associated with the extraction of metals such as iron ore, copper, gold, or nickel, or with crystals such as diamonds, but it also includes the extraction of sand, gravel, clay, ceramic, chemical deposits, fertilizer minerals, and other non-metallic minerals (Spitz and Trudinger 2009, p. 222). The scale of operations is also diverse: there are large-scale surface and underground operations operated by multinational corporations operating across the globe including Canada, Australia, Papua New Guinea, and Africa, as well as some of the new frontier zones including Mongolia, Siberia, and Greenland. There are also many forms of legal and illegal artisanal mining operations which are concentrated in less affluent countries of the Global South.

The "bad old days" of mining are usually characterized as a time when underground work conditions were unsafe, wages were poor, environmental impact was ignored, and when no reparation was made to Indigenous people on whose land mining often occurred. Most transnational mining companies, particularly those who are signatories to the International Council on Mining and Metals ICMM (2013), acknowledge this history and agree with the need to define and abide by international standards in environmental and social sustainability. Yet as this book goes to press, mining has received negative media attention for the following types of issues: human rights abuses in Papua New Guinea (Human Rights Watch 2011), criminalization of anti-mining protests in South America (e.g., Urkidi 2011, p. 566), and corporations accused of being virtual enclaves within states with, for example, their use of private security forces (e.g., Ferguson 2005, p. 378).

This book is part of a series of lectures entitled "Engineers, Technology and Society," and has arisen from a larger project which considered the issue of "Engineering Education for Social and Environmental Justice." Each of the books arising from this interdisciplinary Australian education project critiques a different part of engineering "service:" energy and water; waste management, and even, rather alarmingly, "war." This contribution, focusing on mining practices, examines the social effect of large-scale surface mining on communities within the historical, political, and cultural life of two regions: Papua New Guinea and Western Australia. Our overall purpose is to show that this type of understanding, based on interdisciplinary knowledge, is vital if we are to avoid casting the debate about mining into unrealistic stereotypes in which those who support mining are seen as slaves to capitalism, and those who oppose it are dismissed as polemical activists with no grasp of economic reality.

There is also widespread agreement that an informed debate needs to occur if engineers are to move beyond the stereotype that all engineers have a narrow problem-focused mentality. This is evident in publications from mainstream engineering institutions (e.g., UNESCO 2010) to those with a more critical self-reflexive perspective on engineering practice (e.g., Baillie and Catalano 2009). The need to move "beyond the technical" has been voiced for some time in engineering circles but this view has gained momentum in recent years and there are increasing calls for engineers to "pay attention to the underlying structures that define and shape their work" (Riley 2012, p. 125).

In another publication we have argued that many mine managers are very aware of the consequences of their work (Armstrong and Baillie 2012) and it is not that they "cease to act as engineers" but instead are beginning to additionally "think like anthropologists." Despite that finding, it would be fair to say that for most mining engineers—who focus on a very small technical part of mineral extraction (rock mechanics, say), or mineral processing (such as software programming)—it is difficult to think beyond those particular concerns when they have no direct contact with the people who are impacted by the larger project.

This book is intended to fill that gap by providing contrasting narratives of how mining affects community life. We demonstrate, through the use of these narratives, that understanding local needs (which are often contradictory) and local history (a changing play between culture, politics, and economy) does not fit comfortably with an audit culture that determines whether a company is "engaged" with community. Carrying out a social impact survey which entitles us to tick the right box in a sustainability audit may produce legitimacy in the corporate sphere but it does not produce the kind of knowledge that ultimately challenges us to think more critically about our practice and how it affects those around us.

INTELLECTUAL FRAMEWORK

This book narrates mining histories about copper and gold mining in Papua New Guinea, and iron ore and gold mining in Western Australia. These provide a stark contrast in terms of landscape, state formations, environmental constraints, the culture of the communities who live around the site, and the historical changes that have taken place within the company with regard to community engagement.

The purpose of these narratives is to examine the social reality of the impact of mining in different regions. The narratives, which have been constructed from qualitative research[1] and secondary sources (discussed more fully below) are shaped by an anthropological approach which is deployed within an engineering and social justice framework.

TAKING AN ANTHROPOLOGICAL APPROACH

The anthropological approach, which we use in this instance, refers to a particular way of understanding social life built upon the following assumptions about the kind of knowledge we want to produce:

- Understanding the local's point of view

  This idea has its roots in what Stocking (1992, p. 353) has called "paradigmatic" moments in the history of anthropological thought—the realization that each society operates according to a different cultural logic and that if we are going to understand other people's behavior, we need to see the world through their eyes and understand how their actions may be constrained by particular histories, power relations, and cultural values. Trying to understand other peoples' motivations and desires—particularly as they relate to the complicated field of mining—can be difficult because intention and effect do not coincide, and because people's desires are not always consistent or static. It is for this reason that anthropological methods are often seen as "a licence to explore the curious, the messy, and the unexpected" (Mills and Ratcliffe 2012, p. 147).

  This way in which we approach this task of uncovering the logic or meaning behind the "messiness" of our own, or other's, lives is accomplished by a variety of means, most famously by the ethnographic method whereby a fieldworker spends a lengthy period of time living amongst a community, whether it is in inner city New York or a village in Papua New Guinea, using participant-observation, as well as qualitative tools such as interviews (structured, semi-structured, or open) and surveys. This type of lengthy fieldwork is often associated with postgraduate research and is considered a rite of passage (Mills and Ratcliffe 2012, p. 153) usually resulting in the publication of book-length ethnographies. Much anthropological research however is more short term, albeit with

---

[1] Funded by the then Australian Learning and Teaching Council (ALTC), now the Office of Teaching and Learning (CG10-1519), "Engineering Education for Environmental & Social Justice."

the same goal, and can also be shaped around particular issues and for particular purposes, and these can be termed "case study" or "extended case study" (see Burawoy 1998).

This book is not ethnographic but it is anthropological in that it is primarily concerned with the core issue of uncovering how people experience and perceive the world (in this case, the world of mining) and with the specificities of that experience. We have used published ethnographic material combined with our own interviews, to convey that experience. We have, therefore, used the term "narrative" rather than "case study" as it more correctly captures the unfolding of different desires, ideals, and goals. Filer and Macintyre, writing about mining in Papua New Guinea, comment that anthropological studies "challenge notions of unified interest or consensus at the local level, revealing ambivalence and contradictions" and that social understanding about the impact of mining is always "inflected by the cultural conceptions of change, wealth, and resources that obtain in a community" (2006, p. 221).

- Understanding the history of specific contexts

The need to pay attention to history is crucial, almost inseparable, from the aims discussed above. Evans-Pritchard, a famous figure in British anthropology, wrote that the importance of history lies not in the chronological relation of events, but in its explanatory value in understanding current social life, in making behavior "sociologically intelligible" (Evans-Pritchard 1950, p. 121). Attitudes toward mining amongst a white community living in a small rural town in Australia or North America are inevitably shaped by the social and economic history of the region, and will inevitably be different to the attitudes of ethnically diverse middle class Australians or North Americans living in urban centers. These are obvious connections, and one does not have to be an anthropologist or an historian to appreciate the importance of context, but translating that appreciation into the public arena of mining and community engagement has significant consequences. Jon Altman, for example, is an anthropologist and economist who has collaborated with Rio Tinto on a research project whose aim was to explain why Rio's Indigenous employment initiatives had not made a greater impact on the social and economic well-being of Aboriginal Australians (Taylor and Scambary 2005, Altman and Martin 2009). Altman remarked that one of the reasons for the "disappointing development outcomes is that mining companies and the Australian state seem to have limited capacity to recognise the deeply-entrenched levels of disadvantage experienced by communities adjacent to remote mines…" (2009, p. 5). Anthropologists are, therefore, interested in how political history has shaped the way we see the world and the need to pay attention to the historical dimension of power relations is evident in Altman's statement above.

• Reflexiveness and critical inquiry

Reflexivity is the constant reflection and questioning which leads to interrogations about the assumptions and meaning of taken-for-granted concepts, or conditions of social life. This has a particular importance when thinking about the social impact of mining which is often assumed, by states and corporations, to be fundamental to progress and growth. Questioning these assumptions is not unique to academics in the social sciences; indeed, local people in many parts of the world—from Boddington in Western Australia, to the lower Fly River in Papua New Guinea—can speak about "progress" or "development" with ironic quotation marks around both terms. It has long been recognized that the extraction of metals and minerals from the earth's crust is a dangerous enterprise with potential harm for both humans and the environment (see Agricola 1912). Mining corporations are, however, private enterprises that are required to maximize profits, particularly when large sums of money are expended in high risk ventures. Whether corporations can satisfactorily solve the problematic tension between these domains is an area of debate: sustainable mining, for an anthropologist like Stuart Kirsch, is an oxymoron (2010) while for lead agencies such as the International Council of Mining and Minerals (ICMM) it is both desirable and achievable. In order to highlight this reflexivity, we will often use quotation marks around core concepts to indicate that they are contested and the subject of debate.

ENGINEERING AND SOCIAL JUSTICE FRAMEWORK

Our purpose in this series of lectures is to produce knowledge that emancipates engineering practice from narrow territorial confines, and provides anthropologists with some sense of the cultural and technical constraints under which engineers operate. Both disciplines can then contribute to the alleviation of particular inequalities that reflect a much larger and urgent question—not whether mining is sustainable, but whether the cultural framework for mining (in which the scale of operations and technological requirements are determined by consumption needs and economic rationalism) is defensible?

We are influenced in our analysis by the work of Ursula Franklin,[2] Emeritus Professor of Engineering at the University of Toronto, Quaker, and activist for socially just engineering. She has also studied the historical context of metallurgy and metallurgical practices. In her publication *The Real World of Technology* (1990) Franklin emphasizes that "technologies are developed and used within a particular social, political and economic context" (1990, p. 57) and describes the contemporary context as one that is overwhelmingly shaped by the discourse of the "production model"

---

[2] Franklin's work is discussed more thoroughly in "Engineering and Society: Working Toward Social Justice" in this series, by Caroline Baillie and George Catalano (2009).

(1990, p. 31) in which rational efficiency and technological change is prescriptively shaped by the desire to benefit the few, rather than the many:

> *"The unchallenged prevalence of the production model in the mindset and political discourse of our time, and the model's misapplication to blatantly inappropriate situations seems to me an indication of just how far technology as practice has modified our culture. The new production-based models and metaphors are already so deeply rooted in our social and emotional fabric that it becomes almost sacrilege to question them"* (Franklin 1990, p. 31).

Franklin proposes that there be some sort of accounting system, which looks, not only at the triple bottom line of economic, social, and environmental impacts of any engineering project *before* it is implemented, but to consider also "who benefits and who pays?" within these. Her concern is very much with those at the "receiving end of technology:" "Whenever someone talks to you about the benefits and costs of a particular project, don't ask 'What benefits?' ask 'Whose benefits and whose costs?'" (Franklin 1990, p. 124).

In regarding this question we are reminded that the positive benefits often quoted for certain engineering projects, particularly mining, are in fact directed toward one part of society and the negative benefits affect another, often voiceless part. We attempt to raise these questions in our current text. We are not, therefore, ignoring the significant benefits that mining has and does bring to our lives and lifestyles, but in order to develop a balanced argument overall, we hope to give voice to the potential negative impacts, which often are seen to act against principles of social justice.

METHODOLOGY AND CHAPTER OUTLINE

The following section sets out the structure of the book with an explanation of how the data was collected for the three narratives contained in Chapters 2 and 3.

Before looking at the particular experience of surface mining in two different regions, we sketch out a historical framework of the political economy of mining, by illustrating how successive changes in mining technology have been shaped by the desire to extract more minerals from deeper beneath the earth's surface within an imperial, colonial, or capitalist framework. This is followed by narratives about the local experience of mining in Papua New Guinea (PNG) and Western Australia.

The data on PNG is based solely on the history of the Ok Tedi copper and gold mine, well known for the disastrous impact of riverine waste disposal. This case comprises a rich array of journal articles by anthropologists, lawyers, engineers, and environmental scientists; the official position of the regional government and Ok Tedi Consortium; and reports from the news media. Mining has always been associated with "development" in PNG both as revenue for government and providing "economic growth" for the isolated regions of the interior. The research material reveals that local people, mining professionals, government, and academics are conflicted about the cost of development, but the majority see "development" as an inevitable trajectory in Papua New Guinea's economic future.

Chapter 3, on Western Australia, compares the experience of Aboriginal people in the Pilbara in the northwest, with white settler Australians in the small town of Boddington in the southwest of the state. The Pilbara is a very different context from PNG where mining coincided with the beginning of self-rule. At that time, the West Australian mining companies excluded Aboriginal people from mining activities, and attitudes only changed after the granting of Land Rights and the moral pressure brought to bear upon corporations to right past wrongs. This material is based on interviews with a range of people working in, or affected by, mining: white settlers (politician, anthropologist, archaeologist, mine manager, engineer, doctor, and community engagement personnel) as well as Aboriginal people (an entrepreneur, a cultural awareness educator, health professionals, Land Rights Council employees). Although there is less anthropological material available (compared to PNG), there was enough historical and anthropological data (from academic journals and books) to supplement our research and thus present a holistic account of how mining companies perceive their role in the Pilbara, and how Aboriginal people view mining activities.

The second part of the chapter examines the impact of mining on white Australians in the country town of Boddington, south of the capital city, Perth. The residents of Boddington are primarily white Australians, whom we also designate as "white settlers" to distinguish them from recent white immigrants and from local Aboriginal people. Interviews in Boddington were conducted with a cross-section of the community including volunteers who worked on the local newspaper, women and men who worked at the mine site, an ex-publican, the owner of the local service station, high school students, retirees, and farmers. Published material on Boddington is even less extensive than that available on the Pilbara. This is due to the fact that for a long while mining was not seen as having such an abiding or negative effect on white settler culture when compared to Aboriginal culture, and research on the way in which mining impacts on mental health and community well-being is therefore a relatively new field (see Tonts et al. 2013 for an overview of research in Western Australia). The Boddington narrative is therefore less historical and more synchronic, capturing a range of community views based entirely on interview material.

The final chapter synthesizes emerging themes in these two narratives. We argue that critical self-reflection is crucial for all engineers, particularly those working in the mining industry where the consequences of intervention are profound. This publication forms part of a larger research project—Engineering Education for Social and Environmental Justice—and the scholars who participated in this project (anthropologists, engineers, historians, educators, philosophers, and lawyers) agreed on this common tenet: that critical understanding of core concepts such as culture, power, or colonialism has transformative possibilities for engineering education, but that, conversely, social science and humanities scholars also need to understand the importance of engineering knowledge and practice in our everyday lives.

This publication fills a unique interdisciplinary space by combining an understanding of mining operations with an understanding of how mining impacts on peoples' lives in different contexts. It is intended to help both engineering and social science students, academics, and teachers begin to question their own assumptions and to frame some appropriate alternative processes of community engagement for contemporary and future mining practices.

CHAPTER 1

Mining in History

Mining is commonly associated with transnational corporations such as the members of the International Council of Mining and Minerals. These include companies such as Rio Tinto, BHP Billiton, AngloGoldAshanti, Barrick Gold, Codelco, Inmet, and Xstrata (ICMM 2013d). A common image which we associate with these companies is of large haul trucks transporting ore around the sides of an open-cut or surfaces mine, often in remote regions.

Figure 1.1: Surface mining, photo provided by Wendy Cumming-Potvin.

Mining has always had its critics and supporters. The use of metals has undoubtedly allowed humans to transform the environment to subsist, produce energy, create shelter, produce status goods, and so forth, but the way in which extraction and production is organized reflects a particular form of economic structure which in turn reflects dominant cultural values. Before we can understand the social impact of mining, we need to understand the historical formation of mining economies and technologies. The aim of this chapter is to take a brief—and very broad—critical look at that history which will help us to understand the financial and technological scale of contemporary mining practice.

1.1   INTRODUCTION: HISTORIOGRAPHY OF MINING

The way in which the history of mining is framed ultimately depends on the position of the author, and the following brief overview reveals an historical tension between grand modernist narratives which chart a progressive association between mining, modernization, and development, and counter-narratives which question the benefit of mining to society.

The archaeological classification of human history into the Stone Age, the Bronze Age, and the Iron Age reflects the way in which technological development, closely associated with the ability to mine and process metals, is associated with the transformation in human organization from simple to complex societies. Gold and copper, which are non-ferrous metallic ores, provide a useful means of illustrating the historical importance of metals in human societies. Copper and gold were among the first metals utilized by humans; too soft to create tools, pure copper was probably first used in ornaments but when smelted with tin ore, creates an alloy—bronze (Lynch 2002, p. 10). Copper and its principal alloys, bronze and brass, were used as decorative embellishment and, because of their ability to resist corrosion, for more functional purposes. Brass did not come into use until Roman times and, until the production of iron, copper alloys were used predominantly for weapons, tools, weights and measures, water pipes, roofing, household utensils, mirrors, razors, and artistic decoration (Cordero and Tarring 1960, p. 26). Many of these goods were utilized predominantly by the Roman nobility.

The properties of copper (malleability, ductility, conductivity, and ability to withstand corrosion) have continued from ancient times to the present day:

"... copper has established numerous crucially important uses in virtually all branches of mature industrial or more newly industrializing economies, most notably construction, transport, telecom, and all kinds of electrical and electronic appliances. Even though substitution in favor of aluminium, plastics and glass fibre is always a threat to copper demand, it is hard to conceive a modern society managing without a large-scale and secure copper supply" (Radetzki 2009, p. 176).

An economic historian, Radetzki takes a deterministic view about the role of metals in shaping human society. He describes copper as "the third most important metal in the service of man" (2009, p. 176) after iron and aluminium. This view is reflected in much older mining histories, such as the classic *De Re Metallica* written by Agricola in 1556 and in conventional mining textbooks such as *Mining and the Environment: From Ore to Metal* (Spitz and Trudinger 2009). In this publication, the authors assert that "without mining there would be no civilization as we know it"

(2009, p. 2). Both authors acknowledge the environmental cost of mineral extraction and processing but claim that the advantages outweigh the costs to environment and societies. This is a common, progressivist view of mining which is often used in defence of the industry, particularly in response to those who (while using objects comprised of mining products) are critical of the negative social impact and environmental degradation associated with the industry.

1.2   MINING PROCEDURES

Base metal mining is now considered a "capital intensive" activity in contemporary society (Spitz and Trudinger 2009, p. 225); that is, it requires a high volume of production and a significant margin of profit to provide adequate returns on investment. A brief explanation of the mining project cycle will provide a framework for understanding historical precedents which have shaped current approaches to mining.

The process of metal mining generally involves the following phases:

• Exploration

In the past, mineral deposits were closer to the earth's surface and thus easier to determine by examining visual clues but it is much harder to detect deposits far beneath the ground. Exploration is now carried out with advanced techniques using aircraft or satellites. This is often carried out by junior, or small mining companies who "explore for mineral deposits with a view to negotiating with larger companies or develop a small ore body themselves with their own resources"; the failure rate of these companies can be high due to the financial risk associated with exploration (Rankin 2011, p. 113 and 117).

• Definition and feasibility

Once mineral deposits are identified, it is then necessary to define the "economically mineable part" of those deposits in which, to be defined as an ore body, the concentration of minerals must be such that it can "be profitably mined and processed with current technology and at current market prices" (Rankin 2011, p. 86). The grade or concentration of an ore mineral, or metal, as well as its form of occurrence, will directly affect the costs associated with mining the ore (Rudenno 2004, p. 6). The depth and the size of an ore body will generally determine how it is mined, that is whether it is extracted through surface mining or underground mining. At the present time, surface mining is more common than underground mining because it is less expensive; usually only high grade ores warrant underground mining (Spitz and Trudinger 2009, p. 191). Engineering and financial models are then used to determine "within a framework of commodity prices and exchange rates, the economic return that can be expected"; these models have also to take into account the development of infrastructure required for the project (Rudenno 2004, p. 6). If these evaluations indicate there are "insurmountable technical difficulties in mining, mineral processing, or waste disposal…insufficient financial returns …or that regulatory conditions are too onerous" then mining projects may be abandoned at this stage.

- Extraction

The extraction of the ore from the ground involves drilling, blasting, and transporting the broken ore and waste to a processing location. Tunneling in waste rock for underground mines, or "stripping" the ore body in open pit mining, generates waste rock piles of significant size.

- Mineral processing

    ○ crushing or grinding it to smaller pieces.

    ○ different techniques such as smelting or leaching are used to separate the concentrates (valuable minerals) from the tailings (unwanted minerals or unmineralized rock). As ore bodies often contain a large percentage of unmineralized rock, the tailings must be stockpiled and managed to avoid environmental problems.

The technical components of each stage have changed over time and are shaped by the culture, economy, and society in which mining takes place.

1.3   SOCIETY, ECONOMY, AND TECHNOLOGY: MINING IN HISTORY

In this section we will consider the broad historical phases in mining within the European context. As humans had to dig deeper to extract these minerals in order to satisfy the expanding needs of increasingly complex societies, the processes became more capital intensive (thus affecting the formation of social classes) and the associated technologies caused greater environmental degradation. Overall, there is a historical trend from artisanal or small-scale mining of surface ores (organized by and within communities) to the more complex excavation of deeper ore bodies with associated impacts on politics, society, environment, and culture.

The earliest method of extracting copper from a surface ore body was smelting, whereby the minerals (containing copper) were heated and the copper ore was reduced into metal, leaving slag as a waste product (Radeztki 2009, p. 178). The production of bronze probably arose from mining surface ores that contained both copper and tin. The Bronze Age in Europe extended from about 3000-600 BC and the forms of mining in this period were "dictated by the ore body" and by technology: "pits rarely penetrated below about 10 metres below the surface" (Craddock 1995, p. 31).

Large-scale mining of surface ore bodies did occur in ancient times, particularly during the Roman Empire (27BC–476AD) when the Roman state sought both utilitarian metals such as iron, copper, tin, and lead, as well as gold and silver (Duncan 1999). This is because Italy, "although rich in iron, could not provide a sufficient supply of the whole range of metals needed by the Roman state for coinage and by members of the elite for the luxury artefacts that helped to enhance their social status" (Edmondson 1989,p. 84). The rights to mine metals and stone were leased by the state to private entrepreneurs who worked the mines and quarries with slave labor (Rostovtzeff 1957, p. 299).

It was not until the late Middle Ages, however, that we see how technological changes were used to mine on a large scale in Europe, thus generating great wealth for the emerging merchant classes.

1.3.1   LIQUATION: MINING AND MERCHANT CAPITALISM IN EARLY RENAISSANCE EUROPE

Copper ores often contain silver and gold. The process of liquation—separating silver from copper—emerged at the beginning of the Renaissance in the 1450s in the town of Nuremberg from where it spread throughout Germany, Poland, and the Italian Alps (Lynch 2002, pp. 20–21). It is here that we see the beginning of mineral processing as a large commercial enterprise:

> *"Essentially it relied on the fact that when a furnace packed with a mixture of lead and silver-bearing copper ore was smelted, in the process of cooling down, the silver contained within the ore would be absorbed by the lead. The result was a congealing mass of molten metal, in which the silver had moved from the copper to the lead. The next step was to separate the lead from the copper"* (Lynch 2002 p. 22).

This was relatively simple to achieve as lead has a lower melting temperature than copper, but it was efficient (i.e., worth buying the lead required) only if it was performed on a large scale; the operation typically took place in a very large building and was not suitable to backyard, small-scale operations. It was also a difficult process, so the management of liquation remained in the hands of expert metal workers.

Figure 1.2: The process of liquation (Agricola 1912, p. 504).

The emergence of new mining technologies with the beginnings of capitalism at this time meant that large amounts of wealth were generated and controlled by small groups of people. The individuals who experimented with mineral processing—alchemists or artisanal workers—were not those who profited most from the mining booms of the fifteenth and sixteenth century. Investment funding came from merchant banking houses, many of them German (Blanchard 1998, p. 7). Mining attracted investors from the merchant and aristocratic classes who were "after quick returns on their money" (Lynch 2002, p. 23). The process of liquation increased the demand for lead. Thus when new deposits of lead were discovered in Poland, this attracted further investors from Germany, Italy, and Sweden who went on to invest their profits in real estate (Postan 1987, p. 559).

The use of liquation enabled old fields to be re-mined and new fields were mined more extensively. The extent of these operations has also been partially gauged from the level of waterborne metal pollution detected in lake sediments in central Sweden, a historically important mining region in Europe. Focusing on lead, a group of scientists traced mining impacts on rivers draining into the Baltic Sea (Bindler et al. 2009). They found evidence of pollution from lead mining in sediment and peat records dating back to the 10th century but with the greatest impact occurring during the 16th century onward.

The only constraints on the commercial mining of ores were environmental: timber was used as a fuel to smelt ore and the availability of timber provided a limit upon the extent to which a mine could be exploited. But the rising cost of precious metals during the 16th century encouraged investors to take a risk with capital intensive operations and to trial new technologies and procedures which were often dangerous to humans and environment (Miskimin 1977, pp. 28–29). The organization of labor at that time also demonstrates the beginning of the commodification of labor before industrialization. "Having long been independent artisans," Braudel claims, "working in small groups, they [the miners] were obliged in the fifteenth and sixteenth centuries to put themselves under the control of the merchants who alone could provide the considerable investment required for equipment to mine deep below the surface" (1982, p. 52).

This brief account demonstrates how one form of mineral processing—liquation—required a large outlay of money to be effective and thus generated wealth for the merchant classes who could afford to invest in it. The first wage labor in mining appeared in Germany, when mine workers in the 15th and 16th centuries offered themselves as laborers to people who owned the machines in the mines (Graulau 2008, p. 133). Much of our knowledge about the technology of mining in the 16th century comes from Agricola, a physician who lived in Joachimsthal, a mining town in Germany. Agricola's manuscript *De Re Metallica* is both a practical description of, and philosophical justification for, the mining practices of his time:

> *"Without doubt, none of the arts is older than agriculture, but that of the metals is not less ancient; in fact they are at least equal and coeval, for no mortal man ever tilled a field without implements. In truth, in all the works of agriculture, as in the other arts, implements are used which are made from metals, or which could not be made without the use of metals; for this reason the metals are of the greatest necessity to man. When an art is so poor that it lacks metals, it is not of much importance, for nothing is made without tools. Besides, of all ways whereby great wealth is acquired by good and honest means, none is more advantageous than mining; for although from fields which are well tilled (not to mention other things) we derive rich yields, yet we obtain richer products from mines; in fact, one mine is often much more beneficial to us than many fields. For this reason we learn from the history of nearly all ages that very many men have been made rich by the mines, and the fortunes of many kings have been much amplified thereby* (Agricola 1912, p. xxv).

Agricola clearly believed that mining—or rather, metals—benefit society in general and just as clearly recognized that mining reaps great profits for elites. In answering the critics of mining, he said "it is not the metals which are to be blamed, but the evil passions of men which become inflamed and ignited or it is due to the blind and impious desires of their minds" (1912, p. 16).

1.3.2   MERCURY AMALGAMATION: MINING AND COLONIZATION IN 16TH-CENTURY LATIN AMERICA

The Spanish invaded the "Americas"—as they were subsequently labeled by the Europeans—in the 16th century and the search for precious metals was one of the driving forces for this process of conquest and colonization (Dore 2000). As a consequence, the mining, processing and transport of metals were at the centre of the social and ecological transformation of colonial Latin America (Robins and Hagan 2012, p. 627).

Potosi is a town in Bolivia, situated on the eastern slopes of the Cordillera de los Frailes on the Andes ranges, roughly 4,000 feet (1,219 m) above sea level. Prior to the arrival of the Spanish in 1545, it is highly likely that Indigenous peoples in the region were aware of the silver contained in the rocks of the *cerro rico* (rich hill) and would have worked open pits with some tunneling. The ore-bearing rock was softened through fire-setting and then removed with hammers, chisels, and crowbars. Indigenous miners had little difficulty in finding 'native' silver but also mined ores containing chemical compounds of silver. In the former case, ores were carried to streams where a large, rocking stone acted as a crushing mill. The crushed ore was floated along a stone-lined channel of running water diverted from the stream and bits of native metal could be removed by hand using a pan or basin. Indigenous smelting could be done within a hole in the ground or with a furnace made out of stone or clay with vents to allow the wind to fan the flames within (Bakewell 1984, p. 15). Most of the ores which were smelted in this way left a mixture of lead and silver and, in order to separate out the lead, the mixture was heated in a perforated crucible through which the lead would leach out (as it has a lower melting temperature than silver).

With the Spanish colonization of South America and subsequent discovery of silver at Potosi, the scale and extraction methods changed drastically. Mining at Potosi shifted from being a local, intermittent, small-scale surface activity to being a continuous, large-scale underground venture. This shift require the use of timber (to construct tunnels), draught animals (to shift ore and transport food), and food for the miners (Dore 2000, p. 6).

The other major change occurred in the extraction of silver from the ore. By the 1570s the use of the patio process which was based on amalgamating pulverized ore with mercury, then smelting it to remove the silver, enabled the Spanish to mine deeper, lower-grade ores. Mercury is however very toxic; nonetheless the combination of a forced labor system with the technique of mercury amalgamation "enabled sustained, industrial-scale refining of abundant lesser-quality ores unsuitable for smelting" (Robins and Hagan 2012, p. 627).

"By the light of torches miners hammered at ore bodies in underground tunnels. In low, crowded and poorly ventilated vaults the temperatures soared. Mitayos spent hours climbing up fragile ladders with heavy sacks of ore, to emerge into snow and freezing temperatures at the surface. They immediately descended into the inferno below to cut and carry another load. Working

alongside mules, men crushed silver and mercury with their bare feet in large vats. Some died quickly of mercury poisoning; others languished with fevers and sores" (Dore 2000, p. 8).

In this abbreviated account of the notorious silver mine at Potosi, we see how the development of new technologies in a colonial context benefited the ruling classes and merchant elites of the colonizers. As Braudel puts it: "The mining world foreshadowed the industrial world with its proletariat" (1982, p. 198).

1.3.3   DRILLS AND DYNAMITE: THE EFFICIENT USE OF LABOR AND CAPITAL IN 19TH-CENTURY COPPER MINES

Historically we have seen how new technologies have improved capacities to mine larger ore bodies, and to separate minerals from ore more efficiently, yet the beneficiaries of these developments were not the many but the few: the merchant classes and the aristocracy. We shall now present a brief account of copper mining in North America to illustrate how mining at depth during the 19th century was based on vertical integration of mineral processing (i.e., extraction and processing was carried out at the site) which required large financial outlays from investors; this in turn generated large business structures. This narrative also illustrates how the use of new technologies in the 19th century was geared to profitability and efficiency rather than the health and well-being of the workers. The bulk of this material comes from Larry Lankton's *Cradle to Grave: Life, Work and Death at the Lake Superior Copper Mines* (1991).

First Nation Indians had mined copper from deposits on the Keweenaw peninsula in Michigan state long before the arrival of European colonizers in the 1600s who lacked the technology to mine and transport the copper. Copper deposits did not attract lone prospectors—unlike the gold rushes in California in the early 19th century—because it required more than "a modest array of tools" to extract copper from the ore (Lankton 1991, p. 8).

This particular ore body contained 2% copper and only a large mining operation could produce enough copper to justify the construction of their own smelters. At this time the mining industry became concentrated in the hands of a small number of investor groups each of which had "struck it rich at least once" (Lankton 1991, p. 21); this reflected the wider trend in North America where many small mining firms were replaced by fewer, larger companies. In many ways, this was the precursor of current day practice whereby mining firms owned the mine, the mill, the smelter and the connecting railroads. The investors lived in urban centers far removed from the mine sites, which were managed by experienced miners (often from Cornwall) overseeing a large pool of immigrant labor.

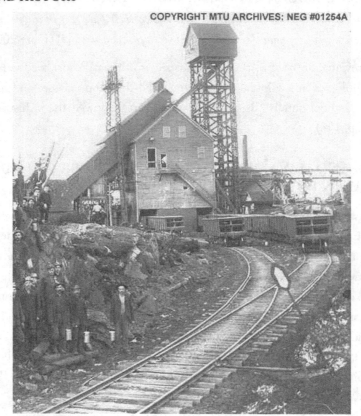

Figure 1.3: Kearsage mine shaft, ca. 1900 (Michigan). Used with permission from Michigan Technological University Archives and Copper Country Historical Collections, http://www.mg.mtu.edu/MINE_SHAFTS/shaft9.htm (Neg#01264A).

The Lake Superior copper mines were below ground, with workers tunneling out from the main shaft along narrow, steep lodes. Until the mid-1860s men moved down the shaft on ladders which affected their health and also affected productivity: mechanical ladders and, eventually, elevator-like cages increased the speed and depth to which men could descend. In underground mining, the void that created with the removal of ore is called a stope (Rudenno 2004, p. 89). "Stoping" techniques have changed over time; in the mid-19th century it described the process of digging out the ore in tunnels radiating out from the main shaft; hand tools were used in the earliest forms of stoping and this technique was improved (that is, it could be done more quickly) with the use of explosives and rock drilling machines from 1880 onward.

Steam power radically transformed the copper mining industry in Michigan after the 1840s. Steam power drove shop machinery, powered generators, drove air compressors, and lifted rock up the mine shafts: overall, it modernized surface operations rather than underground ones.

Copper extraction still remained a labor-intensive industry with large numbers of men required to work underground.

The development of new technology required larger investments in mining operations. The industrialization of the mining economy meant that "it took money to make money" and this fostered a reliance on corporations as America's chief mining promoters (Lankton 1991, p. 58). The expense of drilling and blasting rock, for example, accounted for half the total cost of producing copper in the mid-19th century.

The pressure to recoup money as well as understand the new technology led to changes in mining personnel. Prior to 1900 companies did not employ many university-educated mining or civil engineers. The head mining captain's position was "part manager, part geologist, and part mining engineer," a person with a "nose for copper" (Lankton 1991, pp. 59–60) who had learned the business from the bottom up. The head mining captains and surface captains were usually either settler Americans or Cornishmen who were considered the "elite" of the immigrant workers, but over time "mine management came to be dominated by a new, modern breed of experts who prided themselves on applying science, and not archaic rules of thumb, to the winning of metals" (Lankton 1991, p. 58).

Achieving cost reductions in mining proved difficult as the depth of the mine increased and mine owners turned to the new breed of trained engineers to use "scientific problem solving" regarding the efficiency of all parts of the mining operation. Colleges and universities in the U.S. had taught engineering since the Civil War era: the Columbia School of Mines (1864), the Colorado School of Mines (1874), and the Michigan School of Mines (1886) were the most well known (Lankton 1991, p. 73). By 1910, engineering graduates from these schools became the "efficiency experts" who wanted to improve even the lowliest of mining tasks.

The blasting and extraction of ore was one area where mine management strove to improve efficiency. A two-man drill, a mechanical drill operated by compressed air, was introduced in the 1870s, but by 1913, these were replaced by "one man drills."

The one-man drill was a labor-saving and cost-efficient piece of equipment for the company. However it became known as the widow-maker because of increased injuries and fatalities: "A one-man drill meant there wasn't always a fellow worker nearby, meaning reduced safety when working alone, especially on shaft scaffolds and makeshift board bridges" (Copper Country History, n.d.). This can be seen in Figure 1.4.

The miners of the so-called Copper Counties in Michigan (Houghton, Keweenaw, and Ontonagon) called a strike on July 23, 1913, with three main demands: (1) an increase in their wages, (2) an 8-hour day, and (3) the return of the safer two-man drill (operated by two miners) that had been replaced by a one-man drill (Garrett and Garrett 2013). While Lankton has pointed to other reasons for the strike—such as paternalism, and the tensions generated by ethnic hierarchies among the workers—the introduction of labor-saving devices were seen as threatening, often unsafe, and "made hard jobs harder than before" (Lankton 1991, p. 76).

Figure 1.4: A miner using a one-man drill in one of the Calumet and Hecla underground mines, Calumet, Mich. Photo circa 1915, used with permission from Technological University Archives and Copper Country Historical Collections: http://coppercountry.wordpress.com/tag/one-man-drill/.

1.4   MINING IN THE 21ST CENTURY

Mining corporations now wish to avoid the excesses of the past and to conduct their business in a more sustainable and ethical way. Mining companies cannot, and do not wish to, alienate labor on their sites; nor can they ignore the interests of those people on whose land they wish to mine. They continue to face similar challenges such as declining ore grades of variable quality and purity; increased waste rock; and environmental constraints such as the supply of energy and water to remote sites (Mudd 2010). Unlike their 19th century counterparts, however, contemporary companies are now required to exercise "corporate social responsibility." Companies who wish to become members of the International Council of Mining and Minerals, for example, must agree to:

- Implement and maintain ethical business practices and sound systems of corporate governance;

- Integrate sustainable development considerations within the corporate decision-making process;

- Uphold fundamental human rights and respect cultures, customs, and values in dealings with employees and others who are affected by their activities;

- Implement risk management strategies based on valid data and sound science;

- Seek continual improvement of health and safety performance;

- Seek continual improvement of environmental performance;

- Contribute to conservation of biodiversity and integrated approaches to land use planning;

- Facilitate and encourage responsible product design, use, re-use, recycling, and disposal of their products;

- Contribute to the social, economic, and institutional development of the communities in which they operate; and

- Implement effective and transparent engagement, communication, and independently verified reporting arrangements with their stakeholders.

Commitment to these principles, however, can be put under pressure from three different domains.

Firstly, the requirement to generate a profit is proportional to the financial risk in mining, which is greater than ever because of the size of the investment required in exploration and development of modern mines, and the increasing level of technological complexity required to mine a low-grade ore body. Concentrations of metals such as copper and gold are measured as the proportions in an overall deposit. Copper is sometimes measures as a few parts per hundred (eg 3% Cu) while gold is measured in parts per million (e.g., 6 grams/tonne gold) (Mining Journal 20114). The desire to generate profits as soon as possible to cover the capital invested is thus strong and can, as we shall see in Chapter 2, may generate unethical or unwise decisions.

Secondly, there are serious challenges facing companies which mine low-grade ore bodies, particularly in the areas of energy consumption, water consumption, cyanide use (for gold processing), and waste management (Mudd 2007). Water usage in mining is predominantly used in chemical processing, particularly hydrometallugical processing and has major impact on local and regional use. Where mining occurs in remote, arid regions, it will affect the "flow pattern of groundwater through aquifers and affect water availability from bores elsewhere" (Rankin 2011, p. 216). The extent of waste rock is now likely to be at least equivalent to the amount of ore mined and for many minerals is up to several times higher. Mudd reports that "while a major proportion of this waste rock is likely to be relatively chemically benign, a major quantity is likely to present challenges during operations and rehabilitation due to sulfides present, climate regimes, sensitive environments or communities being adjacent or a combination of these factors" (2010, p. 110).

Finally, commitment to the concept of "social responsibility" is also inhibited by deeper misunderstandings about local culture and economy, and the ways these intersect with the values, not of the mining company, but with the values of the economy it represents. The reality of mining, and the lived experience of those who live alongside mine sites, does not, therefore, always match the commitment to the ten principles listed above. In the following chapters we will look at this disparity more closely by examining the complex engagement between mines and communities in three different settings: the remote tropical highlands of Papua New Guinea; the remote desert regions of northwest Australia; and the rural hinterland of Perth, in southwest Australia.

CHAPTER 2

The Ok Tedi Mine in Papua New Guinea

2.1   INTRODUCTION

Mining in Papua New Guinea (PNG) has had a troubled history, and many current operations continue to attract the critical interest of a range of local and international non-government organizations.[3] The Bougainville copper mine is the most notorious case in that history; discovery of copper deposits in the 1960s led to the establishment of a huge copper mine on the island of Bougainville which was jointly owned by Conzinc Rio Tinto Australia (CRA) and the PNG government. At that time, it was the largest open cut mine in the world. It brought significant revenue to the government but also caused major environmental degradation and social conflict: thousands of immigrant workers from other parts of PNG came to the island "bringing crime and alcohol abuse with them," rivers were polluted, and local subsistence activities could not therefore be maintained (Ross 2004, p. 24).[4] Local groups formed the Bougainville Revolutionary Army in 1988, demanding both the closure of the mine and independence from the PNG government. Their attacks on the mine forced its closure in 1989 but the violence continued until 1997, during which time almost 10,000 civilians died (Ross 2004, pp. 24–5). Academic commentators emphasized that environmental degradation was not the sole cause of the Bougainville conflict: some of it was due to political tension but much of it was due to a lack of consultation with local groups and a fundamental lack of recognition that cash compensation was inadequate recompense for the destruction of subsistence economies which, by extension, shaped social identity and cultural values (Hilson 2006, pp. 27–28, Ogan 1991).

---

[3] There is a large range of NGO groups acting as "watchdogs" on mining activities in Papua New Guinea. Local examples include ActNow! (2013), Papua New Guinea Mine Watch (2013), and The Watut Cries blog spot (2012), while Cardiff et al. (2012) typifies the kind of international NGO publications on mining.

[4] For background on the Bougainville conflict, see Filer (1990), Regan (1998), Filer and Macintyre (2006).

Figure 2.1: The OK Tedi mine. Used with permission from Ok Tedi Mining Limited.

The Ok Tedi mine also attracted international media interest because the practice of riverine waste disposal caused extensive environmental damage to the Fly River and ultimately led to a lawsuit against BHP Billiton (the major shareholder in the mining consortium) in 1994. In this chapter, we will consider the history of the Ok Tedi mine from multiple perspectives: that of the state, the company, and the communities affected by the mine. This history sheds light on a range of problematic issues, which are still relevant to current mining operations. These include the ability of newly formed or impoverished states to effectively negotiate with, and regulate, mining activity and to act as shareholders in, and regulators of, the mining industry. This narrative also reveals how apparently irrational decisions to continue riverine waste disposal are justified against the need to recoup considerable financial investment (for the company) and to generate revenue (for a developing state such as PNG). Finally we also consider the range of community perspectives on the Ok Tedi mine from those living nearby to those living downriver and how the diversity of views amongst these different groups makes it is impossible to speak of a single "community" when imagining the social impact of mining.

The Ok Tedi story is also important because it is part of the historical backdrop against which the concept of "corporate social responsibility" developed. As we shall see in the following chapter, the idea of needing a "social license to operate" developed, not just out of the land rights movement and pressure from human rights groups, but also on the desire to avoid "another Bougainville" or "another Ok Tedi."

2.2   DESCRIPTION OF THE OK TEDI MINE

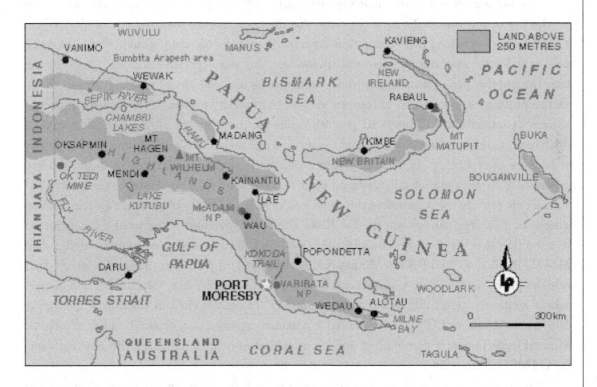

Figure 2.2: Map 1: The Ok Tedi mine, PNG. Retrieved from: http://www.uwec.edu/jolhm/EH/ Eslinger/background1.htm.

The mine site is situated on Mt. Fubilan in the Star Mountains of PNG's western province. The site at Mt. Fubilan—no longer a mountain since mining began—is generally known as "Ok Tedi" because it is situated on the Tedi River (ok: river), a tributary of the Fly River which flows into the Gulf of Papua (see map above). We will use the same terminology to refer to the mine site and its operations, as distinct from OTML (the Ok Tedi Mining Limited) which is the conglomerate which owns and manages the mine.

When OTML began mining in 1984, it comprised BHP Billiton as majority shareholder as well as Inmet (a Canadian mining company) and the PNG government. OTML mined and processed the gold enriched cap of Mt. Fubilan from 1984 to 1986, the extraction of both gold and copper from the copper-rich base from 1986 to 1988 and the extraction of largely copper (with small quantities of gold and silver) from 1988 to the present (MPI 2013). BHP Billiton withdrew from OTML in January 2002 for reasons which are set out below. Since Ok Tedi began operations it has become the largest producer of copper concentrate for the world smelting market, and it contributes a significant amount to the PNG economy in the form of export income, royalties, and taxation payments. It has also been the subject of great controversy.

The mining conglomerate, the PNG national, the regional government, and many local people who live in and around the mine site claim that life has improved since the mine began. When OTML began operations there was a clear expectation that the company would provide financial benefits and other opportunities for local landowners; this was expected—and regulated—by the state. During this period mining companies recognized some ethical and voluntary approach "to practice philanthropy or at least to compensate groups in society for the direct negative impact of business" (Coronado and Fallon 2010, p. 669). While the company has provided medical services which have decreased infant mortality and the incidence of malaria while increasing life expectancy by 20 years (Kay 1995, p. 1), there has been a significant environmental cost to those who live downriver from the mine. A tailings dam,[5] built to accommodate mine waste, collapsed in 1984 immediately prior to the commencement of mine operations. OTML proposed an Interim Tailing Scheme to discharge tailings into the Ok River. The PNG government approved this scheme on the condition that the company consider other sites for the construction of another tailings dam (MMSD 2002, p. 8). In 1988 the Mining Act was modified, in particular the section which dealt with the provision for waste management. In this modified agreement, the construction of permanent waste retention facilities was deferred until 1990 to enable the construction of the copper plant; in the meantime OTML was required to conduct ongoing environmental studies to monitor sediment levels to ensure that "mining operations did not cause unacceptable environmental damage" (MMSD 2002, p. 8).

The rationale for this decision is considered more fully below. In any event, all waste and sediment were discharged directly into the Ok River. By the late 1990s, risk assessment studies (commissioned by OTML in conjunction with the World Bank) confirmed that the 58 million tons of fine tailings and coarse rock being discharged into the Ok Tedi per annum had caused significant environmental damage, including:

- sedimentation which causes overbank flooding and deposition of sediment on the floodplains;

- water logging and sedimentation which reduces the oxygen levels in the soil which will harm or kill vegetation (dieback); and

[5] Tailings are the finely ground material that remains after the valuable metal has been extracted. The tailings from the Ok Tedi mine are composed of fine-grained rock containing traces of copper sulphide and residual cyanide (UNEP 2013).

- decrease in fish stocks due to the increased copper concentration in the water and/or loss of spawning habitations by sedimentation (MMSD 2002, pp. 11–12).

However, downriver residents had been complaining long before these official results were released. In the late 1980s they said that trees along the river banks were dying and vegetable gardens were being destroyed by rising river banks (Kirsch 2002, p. 305). The situation at Ok Tedi began to attract international interest in the early 1990s when Minewatch (an NGO based in London) partnered with the Wau Ecology Insitute of PNG in order to fly several local plaintiffs to testify against BH at the 1992 International Water Tribunal in Amsterdam; furthermore the provincial government sponsored a delegation to the 1992 Earth Summit at Rio de Janeiro "where a Yonggom speaker addressed the media during a press conference held on board the Greenpeace ship Rainbow Warrior II in Rio harbour" (Kirsch 2007, p. 306).

In 1994, ten years after operations commenced, downriver landowners commenced a lawsuit against BHP Billiton, the major shareholder in the Ok Tedi Mining company (Kirsch 2002). Writs were filed against BHP and OTML in the Supreme Court of Victoria, Australia. More detailed aspects of this court case are set out later in the chapter (see Section 2.6). As part of the out-of-court settlement that was reached in 1996 (MMSD 2002 p. 17, MPI 2013) a substantial dredging operation was put in place and efforts were made to rehabilitate the site around the mine (MMSD 2002, p. 15). BHP was granted legal indemnity from future mine related damages through the Compensation (Prohibition of Foreign Legal Proceedings) Act 1995, and in the meantime the OTML consortium continued to monitor sedimentation, copper levels, and acid rock drainage levels.

In 2000 the newly appointed CEO of BHP then declared that, since these levels were not improving, that the Ok Tedi constituted a "dysfunctional aspect" of its portfolio (ABC 2010):

> *"We are not comfortable with our role as operator of that mine and ... we have put together a proposal that we'll be reviewing with our partners and with the Government and other people as to where we might go forward in the future under the assumption that the mine does continue to operate"* (ABC 2010).

The decision to withdraw caused further dismay among the downriver locals and further negative press for the company. Many Yonggom leaders demanded that BHP clean up the mess that it had created (Hyndman 2001, pp. 45–46). It is for these reasons that Ok Tedi is considered an "icon of irresponsible riverine disposal of tailings and waste rock from large-scale mining" (Townsend and Townsend 2004, p. 1).

2.3   MINING GOLD AND COPPER AT OK TEDI

2.3.1   EXPLORATION AND FEASIBILITY: INITIAL INVESTMENTS

This mountainous region (1800m above sea level) is characterized by extremely high rainfall (exceeding 8m per year), heavy cloud cover, and is also prone to landslides and seismic activity (Pintz 1984, 16–19; Townsend 1988; MPI 2013). Even though the area around the mine site is characterized by low seismic activity, earthquakes have nevertheless previously triggered "catastrophic landslides and rock avalanches" (Hearn 1995, p. 47). It has been described as "arduous" (Hettler et al. 1997, p. 280) and "especially harsh by world engineering standards" (Hearn 1995, p. 47).

A prospecting application to explore the Star Mountains was filed by the Kennecott Copper Company (KCC) in mid-1968 (Pintz 1984, p. 32). At that time Papua New Guinea was a Protectorate of Australia. The exploration program, which lasted two years (1969–71) and cost 13 million dollars (Pintz 1984, p. 33) indicated a large body of ore with a high percentage of copper (0.88%). The geologists concluded that "an extremely ambitious engineering project" could convert Mt. Fubilan's 187 million tons of copper ore into a profitable mining project (Hydnman 1994, p. 210).

Kennecott commissioned a U.S. company to undertake an engineering evaluation of the project cost; this estimate came to $AUS339 million (Pintz 1984, p. 34). These high costs were determined, not just by technological requirements of mineral processing but by the cost of implementing that infrastructure in a remote mountainous area with high rainfall and unstable mountain terrain. The Kennecott Company withdrew from the project in 1975 after what has been described as "protracted negotiations" with the newly independent PNG government (Kay 1995, p. 5).

A new consortium was formed with BHP Billiton in 1976. The new Mining Act of 1976 required that mine developers conduct a program of environmental studies; as Townsend and Townsend (2004 p. 4) point out, it is "difficult to imagine how forward looking this requirement was, but at the time this was the most rigorous environmental standard for a mine in a developing country." Government concern about the minimization of social and environmental impact was no doubt fueled by the violent social upheaval and environmental disaster surrounding the Bougainville mine. However, BHP's environmental impact data collection only related to hydrology, water chemistry, sediment transport, landslides, and bird life; it was "optimistically" assumed that waste rock and sediments would wash through the river system without altering the river bed (Townsend and Townsend 2004, p. 5). The government carried out its own studies and it was from these two sets of data that a full EIS was carried out by U.S. consultants which took two years and was submitted in 1982 (Townsend and Townsend 2004, p. 7).

This seven-volume study, produced by a U.S. company (Maunsell and Partners) contained scientific data on water quality, waste rock and sediment issues, the relationship between people, plants, and animals, aquatic biology, and the biological effects of heavy metals, cyanide, and suspended solids (Townsend and Townsend 2004, p. 7). The EIS made it very clear that the impact of the mine would be long term, with river bed aggradation reaching a maximum level fifty years after the mine closed. These findings were never made public and certainly not discussed with local people. Most local people assumed that "the cessation of mining would produce a return to pre-mining conditions" (Townsend and Townsend 2004, p. 6). The authors of the EIS included an anthropologist, biologists, and hydrologists.

The Social Impact Study, on the other hand, was completed in 1980 and was written by a geographer (Richard Jackson) and two postgraduate students—an economist and an anthropologist. This study has since been criticized for being flawed in two ways: it only focused on the immediate vicinity of the mine (assuming that people far down river would not be affected); and it focused on business opportunity rather than local subsistence. Focused as it was on economics, it did not include a cultural heritage component and underplayed anthropological knowledge about the value and social importance of local economies (Townsend and Townsend 2004, p. 8).

Construction had already begun, however, before these studies were released.

2.3.2   MINING AND MINERAL PROCESSING AT OK TEDI

As previously mentioned, the open pit mine on Mt. Fubilan is located near the headwaters of the Ok Tedi in the Star Mountains. The Ok Tedi, along with the adjacent Fly and Strickland Rivers, drain into Fly River catchment. The Strickland River is not affected by mining operations.

The Mt. Fubilan orebody is covered with a gold gossan cap which is the upper layer of the mountain from which the copper has been leached out due to weathering over a long period of time (Hettler et al. 1997). Open pit mining techniques were used to mine the gossan cap, which had a ppm of 2.4 grams of gold per tonne, during the first four years of the mine operations—1984 to 1988 (MMSD 2002, p. 6). OTML has since focused mine operations on separating copper (with small quantities of gold and silver) from the ore body (MPI 2013).

A description of the current mining operations is set out on the OTML website (2008) but to summarize here, there are seven stages in the open cut mining process at Mt. Fubilan:

1. drilling, blasting, and excavating;

2. using haul trucks to dispose of "dead" ore;

3. using haul trucks to transport ore containing copper and gold to the crusher;

4. transporting ground ore to the mill processing plant;

5. separating copper and gold using flotation devices;

6. transporting slurry (the thickened concentrate) to Kiunga via a 50km pipeline;

7. and finally transporting dried concentrate via barge to the coast where tankers transport the ore to smelters in Japan, for example, or the Philippines.

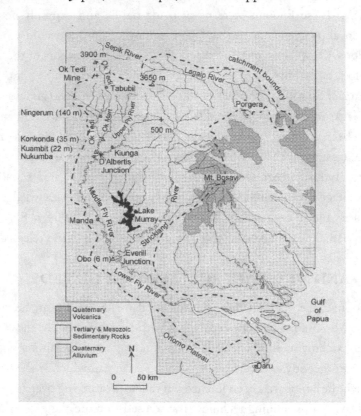

Figure 2.3: Reprinted from *The Fly River, Papua New Guinea: Environmental Studies in an Impacted Tropical River System*, Geoff Pickup and Andrew Marshall, Chapter 1: Geomorphology, Hydrology and Climate of the Fly River System, pg. 5, Copyright (2009), with permission from Elsevier.

As discussed in the first chapter, there are many hazardous environmental impacts associated with mining operations. OTML faced many environmental constraints, such as an unstable terrain and high rainfall, even in the first stages of mining. Slope failures occurred in the immediate vicinity of the mine operation, and in August 1989 a rock avalanche occurred which displaced large amounts of limestone in the vicinity of the mine (Hearn 1995, p. 47). Investigations were subsequently carried out into the relative stability of slopes and creeks in the mine project area with the aim of avoiding further disruptions to mine operations (Hearn 1995, p. 47).

These environmental constraints also impacted on the construction of a tailings dam. The need for such a dam was imperative, particularly for the copper processing which had not yet commenced. Construction began on the dam in 1981 but it was destroyed by a landslide in 1984. There are different emphases in explanations for OTML decision making after this point. According to a retrospective assessment by the Constitutional and Law Reform Commission of Papua New Guinea, OTML argued that, because the area was subject to frequent landslides, high rainfall, and seismic activity, the storage of tailings and other wastes could not be guaranteed, and therefore, the building of a retention dam was too big a risk; although the PNG government did not initially accept this argument, it allowed mining to proceed using an interim tailings dam system (Kalinoe 2008, p. 11). According to an American engineer, building another tailings dam was possible but expensive: three different engineering firms advised that there were workable alternative sites for a tailings dam but OTML "judged their cost to be too high and proceeded to look for a cheaper solution" (Townsend and Townsend 2004, p. 12).

An edited volume of articles was published in 1988 (under the auspices of the United Nations Environmental Program) to assess the potential impact of mining on the Fly River (Pernetta 1988). In the introduction to that volume, the editor stated that "undoubtedly the failure to construct a tailings dam has contributed to the low cost of production" for OTML (Pernetta 1988, p. 6). Bill Townsend, a structural engineer employed by the Department of Mines in the PNG government published a report on the potential impact of riverine disposal in the same volume (1988). Twelve years later, when the Australian Broadcasting Commission produced a documentary, "After the Gold Rush" (ABC 2000) on the Ok Tedi mine, he said:

> *"I got the impression they* [referring to OTML] *were looking for short term solutions not a long term solution, that they were—it looked as if they were postponing the tailings dam to where they would eventually not have to build it all…In order to get back on schedule they pulled crew off the Ok Ma and put them on process-plant construction to get the process-plant construction back on line, which meant that the Ok Ma Tailings dam was getting farther and farther behind … OTML's reaction to spending $300 million on a tailings dam was, "this produces a cost of tailings disposal which is higher than the world standards therefore we will keep looking for another site"* (ABC 2010).

The pressure to start recouping costs from copper mining, plus the cost of building a new dam, forced the cost-cutting measures. There is no doubt that people at that time—both within and outside the mining company—were aware of the potential impact of mining on the Fly River system (Pernetta 1988). The only insight we have on how this issue may have affected those working for BHP at the time comes from the ABC documentary referred to above. In the documentary, Mike Abramski, the chief environmental scientist for OTML until 1985, discussed the time he was asked by the company to calculate the impact of riverine disposal.

> *"We kind of assumed that it was just for comparison purposes, to demonstrate the effectiveness of the tailings dam. So we did some calculations and of course the figures were sky high and ridiculous in comparison to any environmental guidelines"* (ABC 2010).

When he saw the tailings being pumped into the river, he commented that he felt "shock, absolute shock." Mike Abramksi left OTML shortly thereafter.

The Yonggom, with the assistance of local and international activists, initiated a lawsuit against BHP in 1994. Although BHP attempted to halt proceedings, the lawsuit continued and came under extensive international media coverage and unfavorable publicity for BHP. This unfavorable publicity came, not just from the ongoing riverine disposal, but from the subsequent actions by BHP in response to the court action, two of which are cited below as examples:

- In September BHP began a TV, radio, and newspaper advertising campaign with "claims that the tailings released into the Ok Tedi river were 'virtually identical' to natural materials that found their way into the river; that only 20 km of the river's 1000 km were affected; that fish 'seem to be increasing again'; and of the social benefits from the mine (accompanied by TV images of healthy smiling children)" (MMSD 2002, pp. 20–21)

- In the following year, Justice Cummins of Supreme Court of Victoria found BHP in contempt of court for interfering with the administration of justice in Victoria by cooperating with the PNG government in drafting the Ok Tedi Eighth Supplemental Agreement legislation, which would make criminals in PNG of anyone suing BHP in Victoria (MMSD 2002, pp. 20–21).

An out-of-court settlement was announced in 1996. This included a total payment of $110 million Kina (PNG currency) and a further $40 million Kina to the worst-affected areas subject to the conditions imposed on the mining company (a commitment to put an end to tailings from the mine polluting the river; dredging to be put in place immediately) and on the Yonggom people (withdrawal of court action in Victoria and agreement not to pursue further action against BHP) (Kay 1995, p. 7; Kirsch 2006, pp. 17–18; MMSD 2002, 20–21).

In 1997, OTML appointed a group to provide advice, recommendations, and peer review related to a human and ecological risk assessment (HERA) of the terrestrial and aquatic ecosystems of the Ok Tedi/Fly River systems downstream of the mine.[6] At the same time OTML implemented a two-year dredging trial in the lower Ok Tedi to assess whether this would be effective in reducing sedimentation. This dredging program (which cost OTML $30 million per year) lowered sections of the river bed and reduced flooding but only removed about half of the tailings produced by the mine (Kirsch 2007, p. 308). Various options were then considered by OTML to solve the problem of waste management. These included: continuing the dredging program; dredging and installing a tailings pipeline to the take the waste to a land-based storage facility; or closing the mine in 2000. Meanwhile the peer review group had produced a series of reports (Chapman et al. 1997) culminating in a final statement on key issues; in this document they stated that the Fly

---

[6] This group comprised academics from the University of North Vancouver, University of Quebec, Monash University, University of Technology Sydney, and University of California, Berkeley.

River system "will continue to be massively altered for many decades into the future no matter what option (closure to dredging) is taken" (Chapman et al. 2000, p. 3).

Mine closure was not attractive to the PNG government who relied heavily on OTML revenue, nor was it attractive to local government and upriver communities who wished to continue receiving the benefits of the OTML development package. As a compromise, BHP Billiton then proposed to the PNG government that the OTML work toward mine closure by 2005, five years earlier than planned. This plan was also rejected by the government, which preferred the mine to continue for its "maximum economic life" (MMSD 2002, p. 3). At this point, BHP Billiton decided to withdraw from the project.

A timeline of the events up to this point is set out in the table below, which sets out two different perspectives—one largely based on the official website of OTML (2013), and the other based on information from Slater and Gordon, the Australian law firm that represented the Yonggom people (MMSD 2002, p. 23) and from Stuart Kirsch's publication about the Yonggom campaign (Kirsch 1993, 1995, 2007).

Table 2.1: Timeline of events		
	OTML timeline (company perspective)	**Litigation history (plaintiffs perspective)**
1975	• PNG becomes independant	
1976	• The government passes the Mining (Ok Tedi Agreement) Act, and an international consortium is formed to assess the feasibility of developing a gold and copper mining operation	• BHP buys rights to Ok Tedi
1979	• A ten-volume feasibility study is prepared and presented to the PNG government	
1980	• The PNG government approves the consortium's proposals for the project and exercises its option to take up a 20 percent shareholding	

1981	• February 27—Ok Tedi Mining Limited (OTML) incorporated to develop and operate the project • Mining lease is granted to OTML • April—construction begins and proves to be a major engineering feat, given the remote location of the site and the unstable terrain • The development program takes almost eight years and costs US$1.4 billion	
1984	• Massive landslides destroy the foundations of the Ok Ma tailings dam during construction; Mine tailings can no longer be stored • Gold production begins, with an interim tailings system	• Series of accidental cyanide releases into the Fly River
1987	• Copper production begins • Sixth Supplemental Agreement environmental studies begin looking at the effects of the sediment on waters and fish of the Fly River	• German and U.S. interests in OTNL withdraw • Landowners serve a petition demanding recognition of the damage
1988	• Mining of gold cap ceases • Permission granted to dump tailings into river	• Letters of complaint about the quality of water issued to OTML Community Liaison officers
1989	• Government sets maximum sediment level that the mine can place in the Fly River • OTML required to monitor sediment effects	• Threats to block the Tabubil-Kiunga Road
1990	• Creation of Fly River Trust to extend royalties and compensation to downriver communities	• Mine briefly closes as 2,000 protest damage to river and demand compensation
1991	• First dividend paid	

1992		• Representatives of the Yonggom testify against BHP at the 1992 International Water Tribunal in Amsterdam which recommended that the mine should contain tailings waste or close down • International NGO's sponsor Yonggom delegates to hold a press conference at the Rio de Janeiro Earth Summit
1993	• OTML ownership restructure announced; BHP's shareholding moves from 30% to 60%, State of Papua New Guinea 20%, and Inmet Mining Corporation 20%	
1994	• Ten years of production • Ok Tedi/Fly River landowners sue OTML and BHP in the Victorian Supreme Court for environmental damages	• Writs filed against BHP and Ok Tedi Mining Ltd in the Supreme Court of Victoria (Australia)
1995	• First company tax paid • BHP begins advertising campaign about the benefits of mining	• Supreme Court of Victoria recognizes it does not have the power to order OTML to build another tailings dam • PNG Parliament passes the Compensation (Prohibition of Foreign Legal Proceedings) Act 1995

1996	• First general compensation payments made • Mine Waste Management Project (MWMP) establishes to investigate options for mitigating environmental impacts • Heads of Agreement for special benefits signed between OTML, the State, and the lower Ok Tedi communities • Litigation settled out-of-court for K150 million in payments to landowners over the remaining life of the mine, and the commitment to the implementation of a feasibility tailings option which is approved by the State	• BHP began modest payments to villagers, but by the end of the month clan leaders representing 31,488 Ok Tedi and Fly people had opted out of the BHP compensation plan
1997	• Government approves dredging trial in lower Ok Tedi area • OTML puts together a peer review group to provide advice, recommendations, and peer review related to a human and ecological risk assessment (HERA) of the terrestrial and aquatic ecosystems of the Ok Tedi/Fly River systems downstream of the mine	
1998	• Dredging trials begin • State agrees to hold 2.5 per cent of its total equity (now 30%) for the benefit of the mine area landowners and 12.5 percent for the benefit of the people of the Western Province	
1999	• OTML reacts to preliminary results from environmental studies on waste from the mine; the results predict the environmental impact of the mine will be significantly greater than previously expected; OTML flags the early closure of the mine as option for dealing with the problem • OTML releases scientific reports on the environmental effects of the Ok Tedi mine	

| 2000 | • State asks World Bank to review findings
• BHP indicates to shareholders it intends to exit the Ok Tedi mine
• PNG government announces community consultation process over the future of the mine given its environmental impacts | |

2.4   REGULATING THE OK TEDI MINE: THE PNG GOVERNMENT

When it became independent in 1975, the PNG parliament enacted laws claiming subsurface minerals for the state but recognizing the customary land tenure of its Indigenous people under common law. The Ok Tedi mine was the first project to be subject to the regulation of the independent state and the PNG government was in a difficult position at that time of wanting to increase its share of mining royalties and rents—which was seen as essential for the country's development—while at the same time creating favorable investment conditions for foreign corporations. Before examining the government regulation of Ok Tedi in more detail, we will first outline a brief outline of PNG history. The Dutch annexed the western half of New Guinea in 1828 while the Germans and British established protectorates in the eastern half in 1884. With the passing of the Papua Act in 1905 British New Guinea was renamed the Territory of Papua and came under Australian control while control of German New Guinea also passed to Australia following a League of Nations mandate in 1920 (Godden et al. 2008, p. 15). Apart from the period of Japanese occupation during WWII, Australia retained controlled of Papua New Guinea (PNG) until independence in 1975 (Nelson 2000, p. 274).

Mineral exploration, particularly for gold, began in the late 19th century but the major phase of mineral exploration and processing began in the early 1970s, which coincided with the period just prior to independence. One of the main issues for the newly formed PNG government, when it began negotiating with BHP, was that it did not repeat the mistakes that were made with the Bougainville mine site. This mine site, as is well known, generated a great deal of violence and armed resistance (Filer 1990, Regan 1998, Filer and Macintyre 2006). It is less well known that Conzinc Rio Tinto had secured a "very generous taxation agreement" with the Australian Administration (Banks 1993, p. 217) and this was seen as "unacceptable" by the PNG government.

The mineral policy framework of the new government, established by collaboration between members of the new government and a team of expatriate advisers, therefore aimed to:

• Improve fiscal policy by

- ° Decreasing the tax concessions granted to foreign-owned companies under colonial rule

- ° Adjusting tax on company profits to reflect the total amount of profits, so that the government could capture profits when mineral commodity prices increased (Filer and Imbun 2009)

- Require environmental impact statements

- Specify that local people directly impacted by the mine would receive preferential treatment in training, employment, and business development programs (Jackson 2003, p. 4).

Despite later events, it is important to understand the significance of their goals, which represented one of the first attempts by a post-colonial government to regulate a powerful mining corporation; developing these regulations did, however, pose some difficulties. An Australian, Sam Pintz, was employed as the national government's chief negotiator in the Ok Tedi agreements with BHP Billiton. He has acknowledged that his "team of lawyers, planners and financial analysts were not particularly equipped to deal with technical questions" (Pintz 1984, p. 63) and focused instead on negotiating the best deal possible for the Papuan government and created a new Agreement which at the time was seen as a "significant step in terms of developing countries regaining control over their own resources" (Banks 1993, p. 317).

Environmental impact studies were deemed to be the responsibility of the mining company while the government took on the responsibility of carrying out social impact studies. The decision to mine the Mt. Fubilan deposit was made before the government had "formulated its own mining legislation" (MMSD 2002, p. 14). But with the government acting as shareholder and resource owner, many felt that landowners and environment were left "considerably under-represented" in these negotiations (Hyndman 1994, p. 214). Pintz, who was part of these negotiations, acknowledged that the government lacked the capacity to follow through on broad social recommendations (Pintz 1984, p. 142).

While the government clearly hoped to benefit from Ok Tedi in terms of increased revenue, it also believed the local people would ultimately benefit. It was hoped that with cash compensation, the landowners—the Wopkaimin—would undergo a transition from a "traditional" to a "modern" society (Pintz 1984, p. 142).

The question of whether mining represents "progress" or "environmental tragedy" has been debated by the PNG government for some time (Kirsch 2007). Even in the early 1990s many locals were concerned that the lessons learned from Bougainville should not be repeated. When anthropologist Stuart Kirsch returned from fieldwork among the Yongomm and reported on the environmental conditions to a seminar at the University of Papua New Guinea, news of the seminar reached the local newspaper under the headline "Ok Tedi a sewer" (Kirsch 1989); according to one observer, this was treated by the mine management as "nuisance academic commentary" (Burton 2000, p. 99).

It is therefore important to remember the political and economic context of government decision-making regarding the Ok Tedi mine. PNG was a newly independent state for whom foreign investment in mining was a way of promoting modernization and development. Mineral exploitation was seen as a "core strategy" of national planning in a post-independent state (Jorgenson 2004, p. 68) and this created pressure on the government to allow OTML to continue operating without a tailings dam."

The difficulty of being both regular and stakeholder was revealed in other ways. There was, for example, a clear alignment of government and corporate interests in the court case against BHP. As we shall see below, once an out-of-court settlement was finalized with the Yonggom, the government decreed that any landowners in PNG taking legal actions in any foreign courts to be a criminal act thus in effect criminalizing further protest against BHP (Low and Gleeson 1998, p. 207).

2.5   COMMUNITIES AFFECTED BY THE MINE

The following description looks at the experiences of those affected by the mine and reveals a variety of opinions about the benefits of mining from a range of local groups—those who own the land on which the mine exists, those living in close proximity to it, and those who live some way downriver from Mt. Fubilan. Often, when talking about social impact of technology, we speak about "impact on the community"—meaning the local community surrounding the factory or mine site. However, the range of people affected by the mine problematizes the notion that mining companies can list "the community" as a single entity under the category of "stakeholder." It also problematizes the idea that the social impact of mining can be neatly slotted in "positive" and "negative" categories. The views of the upriver and downriver communities, based on anthropological accounts, shift on a spectrum between these two categories with a persistent strand of ambivalence underlying all perceptions of mining. The purpose of this section is to provide the knowledge with which we can think about the question of benefits—not just from the perspective of assessing *what* are the benefits of mining—but of understanding *who* benefits from mining and who pays the cost? (Franklin 1990, p. 124). As we shall see, it is not so easy to define "benefits": what may appear as development and progress from one perspective at a certain time may have negative and unforseen effects with the passage of time.

2.5.1   OUTSIDERS' PERCEPTIONS OF THE STAR MOUNTAINS

Before considering these different perspectives, it is important to recollect how this region, and the people within it, was perceived when Kennecott began drilling around Mt. Fubilan. Many European commentators and urban Papuans described the Star Mountains, and Mt. Fubilan in particular, as "remote," "rugged," and "inhospitable" (an "engineering challenge" according to one geologist), while a team from the University of Papua New Guinea, sent to monitor the health of the Wokpaimin, described the headwaters of the Ok Tedi as "the most remote and least developed part of the country" (Lourie et al. 1986, p. 518). These are common terms used by European observors, particularly colonizers, when describing apparently empty spaces which resist settlement, and are equally common amongst post-colonial government attempting to modernize. Yet the Star Mountains were populated by groups of horticulturalists with a functioning subsistence economy based on a detailed and complex knowledge system about local ecology and landscape.

John Burton, an anthropologist and occasional consultant for OTML, used the term *terra nugax* (*nugax*: trifling, worthless, of no consequence) to describe how the PNG government perceived the Star Mountains. He defined terra nugax as lands which are inhabited but whose social and economic systems were considered of little worth in the dominant "modernization and development" paradigm (Burton 1997, p. 29). There are similar assumptions behind the use of the term *terra nullius* (land of no one) which was applied to the Australian context despite habitation by Aboriginal hunter-gatherers. Unlike European colonizers, the PNG government—based on self-rule—recognized Indigenous title to land and the right to compensation, but Burton emphasizes that decisions could "happily be made about (terra *nugax*) with few repercussions" (Burton 1997, p. 30).

The language of absence, deficit, and lack is evident in both European and urban Papuan perceptions of the land and its people. Prior to mining, the area around Mt. Fubilan was occupied by the Wopkaimin, a group of around 800 people who subsisted on horticulture (taro gardens) supplemented by hunting and fishing (Lourie et al. 1986, p. 520). At that time, the area was accessible only by foot or by air. Hyndman, an anthropologist who has worked with the Wopkaimin since the early seventies, claims they managed to subsist in this mountainous area "through a sophisticated and detailed understanding of local biota and environments and through their time-proven, ecologically and culturally adapted management of resources" (Hyndman 1994, p. 203). Although contact with Europeans may have occurred in 1922, the first sustained encounters did not begin till after WWII when Australian colonial authorities established a downriver patrol post (Hyndman 1994, pp. 205–206). While some described local hamlets as "pleasant," the area and the people were not seen as having any economic potential. For example, in 1964, one patrol officer said that the "Wopkaimin are so far removed and devoid of resources that they simply have no potential for development" (cited in Hyndman 1994, p. 208).

2.5.2 SOCIAL IMPACTS ALONGSIDE THE MINE SITE: THE WOPKAIMIN

These perceptions about the Star Mountains changed after another patrol officer noted in 1966 that the rivers appeared to contain copper deposits. A year later, the Kennecott Exploration Company obtained exploration licenses and established their major base camp at Tabubil from which they began prospecting in the Wopkaimin forests. The Wopkaimin leased areas of their land to the company in return for cash and for promises to build schools and a health clinic. Within a year, the use of Western clothing, tobacco, steel axes and knives, and Melanesian Pidgin English had spread rapidly among the Wopkaimin; influenza had also spread, killing about 5–6% of the population (Hyndman 1994, p. 210). The Wopkaimin were the only local group to receive monetary compensation at the time, unlike the downriver Yonggom and Awin.

Test drilling on Mt. Fubilan began in 1969 and Kennecott eventually had a staff of 45 Europeans and 500 PNG nationals, mostly from the Southern Highlands, with only a few Wopkaimin men. By 1971 Kennecott had financed a first aid post and first primary school with room and board in Tabubil for workers and other Wopkaimin (Hyndman 1994 p. 212). The Wopkaimin were eventually relocated to two roadside villages (Hyndman 2001, p. 36).

The Wopkaimin received cash compensation from OTML and increasing numbers (mostly men) worked for the mine and received a cash income, but a small proportion of the community (mostly women) continued to work in the subsistence economy. The continued influx of cash has however affected the subsistence economy and diet; overall, however, the traditional staple of taro has increasingly been replaced by the faster growing sweet potato (*Ipomoea batatas*); and store-bought and imported Western foods became the main sources of energy and protein for those with a cash income (Ulijaszek et. al. 1987).

The mining town of Tabubil was built in the 1980s specifically to service the mine site, and by the 1990s it had a population of approximately 5,000 of whom 2,000 were direct employees of OTML (Jackson 2003). By 2006, the town had expanded further to include a high school, hotel, banks, supermarkets, and an airstrip boasting connections to Port Moresby and Australia and was described as "a busy modern enclave in an otherwise remote setting" (Jorgensen 2006, p. 235).

Many of those currently residing in the town earn a living from supplying goods and services to OTML. In 2000, for example, OTML employed 2,000 people directly in mining but another 3,500 were employed in supplying goods and services to the company and a further unknown number derive a living from supplying goods and services to mining employees (Baxter 2001 cited in Filer and Macintyre 2006, p. 217).

Figure 2.4: Tabubil town 2004 (photo courtesy of Emma Gilberthorpe).

There is evidence that some health issues have improved since the introduction of health services in the region. The first clinical epidemiological study was carried out in 1978 and indications of malaria, anaemia, malnutrition and respiratory and other infectious diseases were recorded (Taukuro 1980). Subsequent studies indicate that "infant mortality has fallen markedly from over 160/1000 to under 50/1000, and the rate of natural increase of the population has doubled over the past 10 years" (Boyle 2002, p. 63). Cases of the parasitic disease known as filariasis were decreased by 70% after a treatment program, introduced by OTML, commenced in 1986 (Schuurkamp 1990). The impact of the mine (in terms of increased calorie intake) over twenty-five years has also contributed to an increase in height, weight, and Body Mass Index (Adhikashi et al. 2011). These latter figures do not, however, necessarily indicate better health and may indeed be linked to increased propensity for cardiovascular disease and diabetes.

Overall, however, as Jorgensen has commented, most of the Min (incorporating the Wopkaimin and Telefolmin), have been *"active and enthusiastic participants, mainly as employees at Tabubil and secondarily as ben-eficiaries of a number of programs directly or indirectly associated with the mine, including vegetable marketing schemes, provision of health and education facilities, and so on. For those most closely associated with the mine—mainly Faiwolmin, Oksapmin, and Telefolmin —min-ing has figured as the chief source of income for people whose development prospects prior to Ok Tedi were nil"* (2006, p. 238).

In the section below, we look at specific impacts on the Telefolmin, who were not considered landowners of the mine site, but who live in close proximity to Mt. Fubilan. The main source for this material is Dan Jorgensen, an anthropologist who has worked among the Telefolmin since 1980.

2.5.3   SOCIAL IMPACTS UPRIVER FROM THE MINE SITE: THE TELEFOLMIN

The Telefolmin, like many other locals, were hired by geological surveys to cut tracks, collect sam-ples, clear helipads, and build camps in the early 1970s (Jorgensen 1981, p. 67; 2006, p. 241); as many as 40% of men in the villages got jobs in mine construction with OTML (wages comparing favorably with school teachers and public servants); many prospered as entrepreneurs. New con-sumption styles gave the Telefolmin the feeling they were finally enjoying the benefits of develop-ment (Jorgensen 2006, p. 243).

The enjoyment of new-found wealth is, however, offset by other factors, including jealousy between different Min groups, and the impact on the particular social and cultural fabric of Tele-folmin society. Many groups in the local area felt that the Wopkaimin had unjustifiably represented themselves as the only landowners of Mt. Fubilan. Many felt they too should have been compen-sated for the use of land over which they felt they exercised collective rights: "as one individual put it 'it is as if they are cooking a pig and the smell is wafting over to us but they won't let us have a bit of it'" (Gilberthorpe 2013, p. 475).

Jorgensen observed that the increased circulation of cash and the absence of men from the villages affected Telfolmin society and culture in two ways: as a result of men working for OTML for many years, women now take on the brunt of subsistence activities, and bridewealth (payment made by a groom or his kin to the kin of the bride) has become monetized, that is, it is only paid in case not in kind. Bridewealth is inflated to such an extent that men who are no longer employed at the mine find it difficult to find the money for bridewealth. Mining has also impacted on local beliefs. Telefolmin villagers increasingly report sightings of an evil bush spirit called Magalim who is the master of all wild things beyond human control (Jorgensen 2006, p. 246). He is usually sighted when things go wrong. When the tailings dam collapsed for example this was believed to be the work of Magalim who was seen near the site in the guise of a huge speckled python. What is startling is that Magalim is now seen as a tall, white male described as "a suspiciously generous

European offering to buy workers drinks" or more frightening, with a glowing eye or without a head. In all these apparitions, Magalim wears the distinctive yellow boots worn by OTML workers (Jorgensen 2006, p. 247). Jorgensen has interpreted these beliefs as representative of deeper, ambivalent feelings about the impact of mining on their society.

2.5.4   SOCIAL IMPACTS DOWNRIVER FROM THE MINE: THE YONGGOM AND THE AWIN

When mining commenced, there were 3,500 Yonggom residing on the west bank of the lower Ok Tedi and groups of Awin living on the eastern side of the river (Jackson 2003, p. 1). The Yonggom villages were of recent origin, created as a result of the Australian government's wish for them to congregate in larger settlements away from the border with Irian Jaya (Jackson 2003, p. 2).[7] At that time, the Yonggom subsisted on fishing, foraging, and sago (Kirsch 1993b). The population of the Yonggom increased at around the time the tailings dam was destroyed by a landslide, due to in-migration from other Yonggom fleeing political troubles across the border in Irian Jaya (Jackson 2003, p. 5).

The first complaints about the state of the river were expressed by the villagers themselves in 1988 in the form of letters to "as many parties as the writers could think of—the District Office, the government Liaison Officer, the OTML General Manager and so on—making little distinction of responsibility" (Burton 1997, p. 41). They made a variety of complaints, ranging from there being less fish in the river, to the taste of the fish and the taste of the water—"it's something like swallowing a rubber" (Burton 1997, p. 41)—to which there was no response. Many of these letters were difficult to understand because the complaints about the water were interspersed with references to clan mythology, *bisnis,* and development. Because of these linguistic and cultural idiosyncrasies, the validity of their claims was lost and the letters failed to instigate any political process whatsoever.

In the following year (1989) the Community Liaison officers from OTML reported that the village councils were complaining about the pollution along the Fly Rivers and threatened to block the Tabubil-Kiunga road if something was not done about it. But it wasn't until 1990, when a demonstration took place, that the company began to take notice.

Stuart Kirsch, an anthropologist, was hired by Unisearch in 1992, under contract to OTML, to write a report on the social impact of the mine on the Yonggom people (1993b). He reported on the visible damage to the river and the land alongside it, which was "particularly valuable to villagers because it is located within easy walking or canoe distance, and because it offered resources not readily available in the rain forest interior" (Kirsch 1993b, p. 27). Kirsch interviewed villagers from four different zones of the Fly River, all of whom recorded the negative impact of riverine dumping on their livelihood, but many within that group also wanted the mine to stay in operation as long

---

[7] Irian Jaya, or West Papua as it is now known, is a province of Indonesia; Ok Tedi is only 18km to the east of the border (Kay 1995, p. 3).

as it could fix the tailings problem "because of the benefits that it brings to the region, both directly and indirectly, including improved infrastructure and transportation, employment opportunities, and dramatically enhanced revenues" (Kirsch 1993b, p. 3). The crucial question, for Kirsch, was whether OTML could engineer an alternative tailings disposal scheme tomorrow and still make a profit (Kirsch 1993b, p. 3).

In a later publication Kirsch stressed that the Yonggom did not want a violent outcome to their complaints: "they want the mine to stop releasing mine wastes directly into the river system and they want compensation for damage already done to the environment. If this does not take place, they think that the mine should be closed because their quality of life is no longer good. They prefer a political rather than a violent solution to the problems caused by the mine" (Kirsch 1997, p. 122). The feeling of helplessness is reflected in the following quote: "They [the Ok Tedi mine] do not know what we are feeling down here. We are hungry and we are not happy with the pollution. We do not want them to shut down the mine, we just want them to build a tailings dam" (Kirsch 1997, p. 122).

Villagers downstream from the mining area began a campaign in the early 1990s against OTML (specifically BHP) because of the ongoing pollution of their rivers. Representatives from affected communities traveled to Europe and the U.S.; they were supported by the U.K.-based group Minewatch and the Wau Ecology Institute of PNG, who brought several activists from the area to testify against BHP at the 1992 International Water Tribunal in Amsterdam which recommended that the mine should contain tailings waste or close down (Kirsch 2007, p. 306). Two years later they filed a $4 billion claim against BHP in Australian courts over the impact of the mine on their livelihood. Melbourne-based law firm Slater and Gordon represented the villagers (71 landholders representing around 30,000 villagers). The Yonggom were ultimately offered a substantial out-of-court compensation package as well as a promise to mitigate the environmental impact of tailings disposal (Kirsch 1997, p. 137). Further action against BHP is prohibited however under the Foreign Legal Proceedings Act 1995. Regarding the final settlement, Kirsch argues that the "agreement was to settle, that is, to accept less than full value of their losses, as long as the Ok Tedi mine honours its commitment to tailings containment" (1997, p. 130).

However, not all Yonggom wanted the dam to close down, as they were in receipt of benefits delivered to the downriver community. In 1990, the company created the Fly River Trust to extend the benefits of royalties to those living downriver from the mine site (MMSD 2002, p. 14). By 1995, US $3 million was paid into the fund annually and this money had facilitated the construction of "130 community halls, 40 classrooms, 2 school libraries, almost 400 solar lights and pumps, more than 600 water tanks, 23 women's clubs and 15 aid posts" (MMSD 2002, p. 14). Furthermore, the general health of the population in the north Fly region has improved enormously. Infant mortality has decreased, life span has increased from 30 years to 50 years; malaria rates have decreased from 70% to less than 10% (Kay 1995, p. 10).

Money has also been invested in cash crops schemes (such as rubber) with the intent that these schemes provide employment past mine closure (MMSD 2002, p. 14). By 2000 the company had already established over 700 apprenticeships and trainee positions for local and non-local people, and has also developed a similar number of scholarships toward education programs (MMSD 2002, p. 14). Local people are well aware of these benefits and this may explain why, while some groups wanted the mine closed, others simply wanted waste management to be improved.

2.6   RECENT DEVELOPMENTS IN PERSPECTIVE

The Ok Tedi mine has had a tumultuous history since the exploration license was first granted to Kennecott Copper Company almost 50 years ago. While this chapter has focused on the social impact of mining at Ok Tedi up to the lawsuit against BHP, it is worth noting the following developments.

- 2002: After BHP Billiton withdrew from OTML in 2002, its shares were transferred to the PNG Sustainable Development Program Ltd which was established that same year with the specific aim of "applying the funds coming from OTML which are assigned for the development of the PNG, in particular the people of the Western Province" (PNGSDP 2013). The equity participants in OTML were: PNG Sustainable Development Program Limited (52%); the State of Papua New Guinea (30%); and the Canadian Inmet Mining Corporation (18%) (BHPB 2002).

- 2004: All claims against the mine—new ones had been lodged by the Yonggom in 2000—were withdrawn from the courts. Export sales in 2004 amounted to $700 million, representing 25% of PNG's total exports that year (Kirsch 2007, p. 315).

- 2006: Environmental monitoring revealed increased levels of copper in the Ok Tedi which led to the implementation of a Mine Waste Tailings Project whereby sulphur content in tailings would be significantly reduced with the safe storage of the resultant pyrite concentrate at Bige. In that same year, OTML "achieved record before-tax profit of 2.7 billion Kina" (OTML 2001).

- 2010: 1,000 local workers strike for a week over expatriate staffers getting better benefits and bonuses (Gridneff 2010).

- 2011: ruptures in slurry pipeline cause suspension of mining for four weeks at a cost of US$180 million or about K447 million in revenue (PNG Mine Watch 2011).

- 2011: Inmet exits the OTML consortium. OTML is thus nationally owned with PNG Sustainable Development Program owning 63.4 percent and the State of PNG with 36.6% (OTML 2001).

Asked to describe Ok Tedi at the beginning of 2013, the newly elected Governor of the Western Province said "for us in the Western Province we see it as a golden goose. It is there to now help us. Ok Tedi has contributed a lot over many years but particularly now, now that the Ok Tedi mine is 100% owned by Papua New Guinea, and a major part of that belongs to people of the Western Province" (Garrett and Garrett 2013). Previous events would seem to contradict this and the Western province has continues to be described as the "least developed" of the PNG provinces, suffering from poor administration (Garrett and Garrett 2013). The PNG government sacked the chairman of the PNG Sustainable Development program in late 2013, cancelling PNGSDP shares, becoming the sole owner of Ok Tedi (Fox 2013).

The mine continues to operate after a feasibility study determined that the life of the mine could be extended until 2022. The expansion will combine two underground mines and one open pit mine.

2.7   MAKING SENSE OF OK TEDI

2.7.1   THE GOVERNMENT'S PERSPECTIVE

The PNG government explicitly sees mining as the path to development for local people as well as a source of revenue. At one point, the monies earned from Ok Tedi constituted almost one-fifth of the national revenue. It is not surprising therefore that OTML should have been seen as the "only agent of development" in the area (MMSD 2002, p. 20). But what are the consequences when the government is both beneficiary and regulator of mining? This was a pressing problem for the new PNG government which hoped that the "revenues derived from two very large and profitable mines, if properly applied to the task of national development, would enable the government and the country to escape their dependence on Australian aid and expertise before those mines had been exhausted" (Filer and Imbun 2009, p. 80).

The PNG government acknowledges that there is an environmental cost in the journey toward creating a "strong" economy. In 1993, "a senior Papua New Guinean executive at the Ok Tedi mine argued that polluting the river was an acceptable trade-off if the citizens of the country were to achieve their goal of wearing shirts and neckties and having office jobs like Australians" (Kirsch 2007, p. 306).

> *"The audience disrupted his speech with catcalls, and a representative from Greenpeace yelled out that the only reason a real Melanesian would need a necktie would be to hang himself. However, no one at the conference objected when the positive contributions of the mine were subsequently described as bringing progress to an area in which local inhabitants were lacking skills and knowledge"* (Kirsch 2007, p. 308).

This shared notion of progress is what binds the government and the mining consortium together. This is reflected in the Memorandum of Understanding (MoU) between the Ok Tedi Fly River Development Program and the government's Sustainable Development Program which states that they "both share a mandate to improve the lives of Western Province people and communities, particularly within the mine affected corridor" (PNGSDP 2013).

According to the current OTML website, the company's goal of sustainable development is that "at mine closure, mine-affected communities are healthy, educated, and economically independent, with secure access to sufficient food" (OTML 2008b). The company also acknowledges, however, that "At its core … OTML remains a mining company, and for these development benefits to be realised, it must continue to operate cost-effectively within the global copper concentrate market" (OTML 2008b).

Development programs associated with the mining industry, particularly in Papua New Guinea, are based on the conventional perception that development is defined, not just by improved health and longevity, but by cash income from either paid employment or individual initiatives in entrepreneurial activities. As Banks has outlined in previous publications, the benefit packages from mining comprise: on-off or continuing compensation, royalties, dividends from a direct equity share in the operation, "spin off" business contracts and employment (Banks 2009, p. 45). The notion of "economic independence" then is somewhat of a fiction as it enmeshes the locals within the market economy which may come at the expense of subsistence activities which are based on very different principles of production and distribution. Furthermore not all communities may wish to become involved in "productive" businesses, enterprises, or even employment (Banks 2009, p. 46). As Gilberthorpe has pointed out, with specific reference to Papua New Guinea, there are often "ideological incompatibilities between localized principles and development discourse" (2013, p. 468).

There is also potential for tension in the distribution of development packages. Tabubil was described earlier in this chapter as an "enclave of modernity" in an otherwise isolated area. Yet for others:

> "Tabubil, … epitomizes an unequal economic and social enclave where consumables arrive on a truck, and free housing, MTV, Hollywood films, alcohol and fast food are readily available to those receiving cash benefits, but restricted for those who do not. Residents grow dependent on, and are influenced by, a modernized environment that is not sustainable beyond the life of the mine and is not guided by long-term development objectives suited to the region and the people living there. Quite simply, when the mine ceases operating, the status of the town as a 'developed hub' will fall into question" (Gilberthorpe 2013, p. 478).

2.7.2   AN ENGINEERING AND ENVIRONMENTAL SCIENCE PERSPECTIVE

There are no critical reports from engineers who worked for the Ok Tedi consortium at the time. This is not surprising as employees would have likely lost their jobs if they had voiced dissent or criticism from within the company. Some criticism came, however, from engineers outside the company. The United National Environment Programme (UNEP) prepared a series of reports for the PNG government as part of their "action plan for the human environment" with a focus on marine pollution. One of these reports, *The Potential Impacts of Mining on the Fly River* (Pernetta, 1988) was prepared by Australian and Papuan academics from the University of Papua New Guinea; it included a contribution from an engineer, Bill Townsend. Townsend was employed by the Department of Minerals and Energy within the PNG government to advise the state on "technical matters and to co-ordinate the activities of the State's technical personnel working on the OTML project" (Pernetta, 1988, p. 107 fn 4). He was critical of OTML on a number of fronts. He accused them, for example, of making specious claims. Many parts of the Environmental Impact Statement, for example, had serious shortcomings because of the limited time span during which information was gathered. But as he pointed out:

> *"An environmentalist will interpret data so that the grey area of uncertainty is considered unacceptable or at least risky. An entrepreneur on the other hand, will be searching for the limit beyond which damage will occur and thus will tend to consider the grey area of uncertainty as acceptable. That is, there is no damage if it cannot be clearly proven"* (Townsend 1988, p. 109).

John Burton, an anthropologist, also criticized the OTML findings about the projected impact of cyanide tailings as "bad science" because it reported copper concentrations as averages rather than as periodic highs and lows.

Townsend argued that the crucial issue was how much a country or company was willing to spend to protect the environment and reflected that while some companies might be motivated solely by money, governments had responsibilities other than profit (Townsend 1988).

Similar criticisms of the company and the government for approving riverine disposal also came from the principal environmental scientist of BHP at the time. Mike Abramksi was interviewed for the ABC documentary on Ok Tedi:

> *"Firstly I think the consensus amongst the environmental specialists working there at the time was, in my opinion, that firstly, no-one in their right mind would go ahead and build a mine like this without a tailings dam. Not even the most moth-eaten mining engineer would be so out of touch with the present, at that time, as to even propose that. And secondly, that no government would allow that"* (ABC 2010).

For both these individuals, the issue is about following proper procedure with regard to mining processes, regardless of cost. They are arguing for transparency on the part of mining companies and greater regulation by the PNG government. If these had been in place, they argue, Yonggom subsistence would not have been affected and the whole disaster could have been averted. It is implicitly assumed that mining—if properly regulated and ethically driven—can bring development to rural Papua New Guineans.

2.7.3   ANTHROPOLOGICAL REPRESENTATIONS OF OK TEDI

Anthropologists were divided in their views about the impact of the Ok Tedi mine. Hyndman, the anthropologist who worked with the Wopkaimin at Mt. Fubilan, acknowledges that the upriver people did not suffer as a result of the Ok Tedi mine. However, he describes the riverine disposal of waste (and the consequences for the Yonggom) as a form of ecocide[8] and argues that mining in general can never lead to sustainable development (Hyndman 2001, p. 36). He uses a liberation ecology framework which views "growth and industrialization as illusions offered by elites to keep Indigenous peoples from promoting appropriate and sustainable alternatives; reject industrial lifeways; support the defence and restoration of commons, and support decision making by local Indigenous populations" (Hyndman 2001, p. 41). Hyndman believes that to take a stand against BHP for their actions or lack of action regarding the tailings dam reflects his "moral commitment to the Melanesian people and to the wider rights of Indigenous peoples and the destruction of the planet" (Hyndman 2001, p. 41).

At the other end of the ideological spectrum is Robert Jackson, an anthropologist who commented that "if a project like Ok Tedi could increase people's life spans and also providing those people were given a greater chance of self-fulfilment by the project, then the destruction of a few kilometres of swamp forest was a small price—a price admittedly—to pay" (Jackson 1998, p. 308).

Another group of anthropologists have refrained from vilifying either the Ok Tedi consortium or the PNG government, and have looked instead at how to assist local people in mitigating against the foreseeable and unforeseeable impacts that will arise from their engagement with mining. An early critique concerned the amount spent, and attention paid, to the social impact of mining activities not just on the immediate community but on all communities in the region. Burton, for example, was frustrated by the early lack of commitment to understanding the social impact of mining:

> *"The mining industry was willing to spend money on monitoring of the physical environment*
> *but unable to understand and draw up policies for the social environment. In Ok Tedi's case,*

---

[8] This refers to high levels of environmental destruction caused by the "technological and bureaucratic structures that define modernity"—a broad category that could include the effects of commercial fishing or Agent Orange in the Vietnam War—and is controversial because of its association with the term "genocide." Overall it is used to denote extreme forms of environmental racism and injustice (Gottlieb 2010).

company expenditure on environmental monitoring in the decade after the opening of the mine amounted to over K50 million, but barely K0.05 million on social monitoring. This amounts to a spending ratio of about 100:1" (Burton 2000, p. 102).

Some in this group, despite—or perhaps because of—working closely with mining companies, are also highly suspicious of what "development" means to mining companies.

2.7.4 THE MINING INDUSTRY

The phrase "mining industry" refers to the management of mining corporations whose interests cannot be assumed to mirror the individual views of engineers and scientists working within or alongside the corporation (see Section 2.7.2). We are concerned here primarily with how the Board of BHP—the major shareholder in OTML at the time—reacted to, and managed, the environmental and social impact of mining activities at Mt. Fubilan. We do not know for example whether the engineers on site approved or disapproved of the decision not to construct another tailings dam but we can assume that the final decision reflected the economic rationale of senior management who wished to recoup considerable financial investments by going ahead without a tailings dam.

BHP reacted to the lawsuit by attempting to have the case thrown out of court and by trying to claim that the Victorian Supreme Court had no jurisdiction to hear the case. These attempts were unsuccessful. What is more troubling is the apparent collusion between BHP lawyers and the PNG government to draft legislation proscribing compensation claims in foreign courts against resource projects in Papua New Guinea. This legislation, known as the *Prohibition of Foreign Legal Proceedings* Act legislation, was successfully passed in 1995 (Independent State of PNG 1995).

The ability of local people to organize a lawsuit against BHP demonstrated both to the government and to other mining companies "the perils of an unswerving commitment to a top-down, large scale, foreign investment led natural resource development path which does not recognize and provide for local and regional input" (Banks 1993, p. 324). The response of BHP Billiton, to the particular issue of Ok Tedi, was to walk away from a mine which had created significant reputational damage.

The history of mines such as Ok Tedi, and Bougainville, have had a significant impact on the larger mining industry in terms of how mines operate, and also on their engagement with local communities: "One of the by-products of the Bougainville rebellion was a much higher level of political and academic interest in the social (as well as the environmental) impact of large-scale mining operations in Papua New Guinea and other parts of Melanesia" (Filer and Macintyre 2006, p. 218). The subsequent court case against BHP, together with international focus on the re-definition of sustainable development (WCED 1987), were part of an overall impetus for mining companies to reassess their strategic goals; in "developed" countries, this translated to a focus on environmental issues while in "developing" countries, companies were perceived as colluding with authoritarian regimes, failing to consider the social impact of their development schemes, or that they were "irresponsible operators" (Danielson 2006, p. 16). These parallel developments were instrumental in the formation of such concepts as "corporate social responsibility" and harnessing these to the platform of sustainability.

The current management of BHP Billiton, for example, is now keen to distance itself from the management decisions of the past:

> *"(BHP's) CEO, Chip Goodyear, has previously gone on the public record, saying that the controversial Ok Tedi copper mine in Papua New Guinea (PNG) would not have been developed under the company charter BHP Billiton works by today ... Ian Wood, the company's vice president, Sustainable Development and Community Relations, readily concedes that the environmental impacts that resulted from the development of the mine are not something of which BHP Billiton is proud.* The decisions that largely set in train the environmental impacts were taken by the project shareholders, including the PNG Government, in 1982. Faced with a similar decision point today, with the benefit of the science as we now know it, it seems certain that we would have achieved a very different outcome,' *he says"* (AICD 2006).

Critics of corporate social responsibility in PNG however see it as a "bid to legitimize the sector after decades of environmental disasters and the trampling of Indigenous rights" in the region (Gilberthorpe and Banks 2012, p. 185). Gilberthorpe and Banks argue that there is a tendency to focus on the environmental aspects of CSR while the "community development programs that form part of the CSR package often remain less sophisticated, focusing on infrastructure (schools, hospitals, roads), job training and micro credit schemes" (2012, p. 186). They believe that PNG is illustrative of the "ill-conceived and highly inappropriate development programs that contribute little to, have a divisive effect on, the social and economic security of local communities" (2012, p. 186).

The reaction of mining companies to ongoing criticism—coming from academics such as Gilberthorpe and Banks (see also Kirsch 2010, Benson and Kirsch 2010) and from national and international NGOs—is diverse, ranging from defensiveness to a recognition from within the industry that "the best practice of today is still not good enough" (Danielson 2002, pp. 22–23). In order to make sense of some of these comments, however, we need to understand the social reality of "best practice" and how ideals about community engagement are experienced in the "vernacular reality" of everyday life (Franklin 1990, p. 36).

The following chapter will set out the way in which two different companies—Rio Tinto (Pilbara Iron) and Newcrest (Boddington Gold)—have interpreted and exercised the idea of "community engagement" within Western Australia, and how affected communities have responded to them.

CHAPTER 3

Mining and Society in Western Australia

Western Australia is the largest state in Australia and the least populated, ranging from the tropical Kimberley region in the far north, across the dry regions of the Western Desert, and down to the cooler reaches of the far south. It is a vast region home to only 2.45 million people (ABS 2013), the majority of whom are Europeans, descendants of white settlers or more recent migrants.

The original inhabitants of the land, the Aboriginal people, constitute only 3.3% of the total population (AIHO 2012), a statistic which reflects Western Australia's colonial history, where the British first planted the Union Jack on the south west coast in 1796 and established the Swan River Colony in 1821 (Green 1981). The distribution of Aboriginal people across the state is significant because the proportion varies according to remoteness from the capital city of Perth. For example, although just under 40% of Aboriginal people live in the Perth metropolitan area, they constitute less than 2% of the metropolitan population; over 40% of Aboriginal people live in regional and remote Western Australia and it is in areas such as these that mining activities take place.

Settlers perceived the west as a frontier, that is, as an untamed land, and the myth of the frontier allowed settlers to subjugate nature and Indigenous people and to legitimate their actions as "natural and inevitable for ensuring the 'progress' of civilisation" (Furniss 2005). Western Australia was perceived as a place where it was possible to create wealth and recreate oneself in terms of a different, even improved, status or class. The Aboriginal experience of settlement, explained more fully below, was the opposite: the reversal of status, the deprivation of livelihood, and social exclusion. Although the Western Australian economy relied heavily on pastoralism and grain in its early history, mining soon began to dominate the political, social, and economic landscape of Western Australia. From the Kalgoorlie gold rushes in the 19th century to the recent boom in iron ore mining in the Pilbara, mining activity has gone through successive periods of boom and bust. This cyclical activity has plateaued in recent times with the sustained demand for iron ore and the recent expansion of gold mining at Boddington just south of Perth.

This chapter will describe different views about mining in two different locations: the Pilbara region and the Boddington mine site in the southwest of the state. These sites provide an interesting comparative perspective. The Pilbara (see Figure 3.1) is situated in the far north of the state and is home to numerous iron ore mines situated in remote regions where there are a number of Aboriginal settlements. The town of Roebourne is situated on the outskirts of Karratha.

Figure 3.1: Western Australian regions. Retrieved from the Business Migration Centre: http://www.businessmigration.wa.gov.au/?page=living-in-the-regions.

A small country town, Boddington is located approximately 150 kilometres southeast of Perth (the capital city of Western Australia) in the Peel region (see above). Comprised of (mostly) white settler Australians, Boddington is situated between a recently expanded gold mine (Newmont) and a bauxite mine (BHP Billiton Worsley Alumina).

3.1   THE PILBARA

3.1.1   RESEARCH MATERIAL

This section is based on a synthesis of the following material: secondary history, ethnographies, and interviews with a group of people who have diverse views about the impact of mining on Aboriginal people. It has been noted that much of Aboriginal history in Australia is not written by Aboriginal people (Brock 2004) and Western Australia is no exception: much of the available historical material is written by white Australian academics albeit with a critical view of the impact of colonization on Indigenous people. Aboriginal people from the Pilbara have, however, published their own accounts of the past in the form of life histories, such as Smith (2002).

Unlike Papua New Guinea, there is comparatively less ethnographic material available on Aboriginal people in the Pilbara. The following publications were of particular relevance to this chapter: the work of Mary Edmunds, a social anthropologist, who has written insightful social histories of the town of Roebourne (Edmunds 1989, 2012a, 2012b) and also the work of Robert Tonkinson whose work on the Mardu of the Western Pilbara—although not the central focus of this chapter—has shed important light on shared central concepts in Aboriginal culture (Tonkinson 1991, 2007a, 2007b).

There is also a body of work on Aboriginal responses to post-colonial social and economic policies. Much of this work is generated out of the Centre for Aboriginal Economic Policy Research (CAEPR) at the Australian National University (ANU) and is available online through the ANU e-press (e.g., O'Faircheallaigh 1995; Holcombe 2004, 2005, 2010). CAEPR also secured government funding to collaborate with Rio Tinto Australia to critically analyze the impact of mining on Aboriginal people in Western Australia, the Northern Territory, and Queensland, and this has resulted in two publications which are relevant to this chapter: Taylor and Scambary's *Indigenous People and the Pilbara Mining Boom: A Baseline for Regional Participation* (2005), and Altman and Martin's *Power, Culture, Economy: Indigenous Australians and Mining* (2009).

We also conducted interviews with a range of people with different views about the impact of mining on Aboriginal society and culture. These include:

- Anthony Hodge, President of the International Council of Mining and Minerals

- Fred Chaney, politician and lawyer, former Federal minister for Aboriginal affairs

- Verna Voss, senior Aboriginal woman from the Eastern Goldfields, cultural awareness educator for mining companies

- Barry Taylor, senior Aboriginal man, CEO of contracting company to mining companies in the Pilbara

- Robert Tonkinson, anthropologist who has worked with the Mardu in the Eastern Pilbara for forty years

- Bruce Harvey, Global Practice Leader, Communities and Social Performance, Rio Tinto

- Janina Gawler, General Manager Communities, Rio Tinto (Pilbara Iron)

- Don Wilson, Anglican priest and former pastoralist, with ties to the Mardu community at Wiluna (Northern Goldfields)

- Male mine manager of an iron ore mine in the Pilbara

- Male FIFO engineer Pilbara

- Retired male engineers from Wiluna (Northern Goldfields)

A minority of these interviewees chose to be anonymous and their interviews are referenced as, for example: (Male FIFO engineers, Interview 2011). Excerpts from the remainder of the interviews are referenced as follows: (Gawler, Interview 2011). These interviews were semi-structured, in that they centered around two core questions: "what is your perception/experience of the impact of mining on Aboriginal people?" and "what do you think engineers need to know if they are to enlarge their understanding about the social impact of mining?" The answers to the second questions are considered in the chapter which follows.

Finally, we chose to focus on Rio Tinto Iron Ore to illustrate how attitudes of Australian mining companies toward "social impact" have changed over the years. This choice was shaped both by the availability of research material (such as the Taylor and Scambary report, mentioned above) and by access to key personnel who agreed to be interviewed for this project.

3.1.2 HISTORICAL SETTING

Typical of other frontier settlements, such as the U.S., West Australian colonial history is marked by European settlement and a system of government which did not recognize Aboriginal rights to land, political participation, or cultural recognition. The British first claimed possession of Western Australia in 1791, when they anchored in King George Sound as part of a voyage of discovery around the southwest corner of the state (Green 1981, p. 74), but it wasn't until 1826 that they established formal political control through establishing the township of Albany.

By the middle of the 19th century, the local economy in the southern part of the state was dominated by pastoralism, and settlement was mainly confined to the coastline. The first expeditions to the north of the state took place in 1858 and the subsequent reports encouraged pastoralists to take up land in the Pilbara between 1863 and 1866 (Bolton 2008, p. 33).

The occupation of the northwest brought the settlers into contact with local Aboriginal people who had lived in this region as hunters and gatherers for at least 40,000 years, a time span now determined by the dating of rock art (Walker 2009, p. 691). Initial contact with explorers was intermittent and Aboriginal people were often used as guides, but as more land was taken over for grazing, relations deteriorated (Walker 2009, p. 691). Many Aboriginal people worked in the pastoral industry, often, but not always, under very poor working conditions. Others were "blackbirded" (a euphemism for forced labor) to work on pearling vessels (Walker 2009).

Colonial policy in pastoral areas was enacted through local Justices of the Peace and by an active police force that largely obeyed the directives of local landowners. Most of these landowners "had contempt for Aborigines whom they had come to regard merely as another resource to be exploited or as wild animals to be broken and tamed" (Green 1981, p. 95). The police responsible for enforcing a system of indentured servitude, which prevailed during the mid-19th century, were accordingly vested with powers of arrest without warrant (Bolton 1981).

State Library of Western Australia

Figure 3.2: Aboriginal prisoners outside Roebourne Gaol, 1896, (reprinted with permission from the State Library of Western Australia) http://trove.nla.gov.au/version/174378611.

Many Aboriginal men were imprisoned for theft of government property, and for violent reprisals against white settlers who had taken their women or killed their relatives (see Green 1981). Opposition to the practice of indenture and of chaining Aboriginal people by the neck in Western Australia (see plate above) came from a few individuals, but as one historian has put it, these individuals were "whistling up a whirlwind that threatened the economy of the north by exposing it for what it was, the enslavement of thousands of men, women and children" (Green 1981, p. 105) and their views did little to change the status quo. Such was the extent of violence against Aboriginal people that even Sir John Forrest, who vetoed the British government's suggestion that one per cent of state revenue be spent on the protection and improvement of Aboriginal people, said: "There can be little doubt from these frequent police reports that a war of extermination, in effect is being waged against these unfortunate blacks in the Kimberley district" (cited in Green 1981, p.

116). It has been reported that Aboriginal prisoners continued to be chained by the neck until as late as 1956 (Ritter 2002).

Sarah Holcombe, a historian, has explained the government willingness to accede to any dominant group that wished to "develop" the region (2005). Even after the Second World War, she claims:

"There was no representative body, or mediating organization, between Aboriginal people and developmental interests related to exploitation of the land. Relations between Aboriginal and non-Aboriginal people initially occurred through the pastoral industry and according to Edmunds are now viewed by Aboriginal people nostalgically as 'when people lived in harmony'. While this view does not accord with the realities of the harsh indenture system that led to the strike action of the 1940s, it contrasts with the subsequent negative relations with the mining industry from the 1960s and which are now beginning to be addressed" (Holcombe 2005, p. 110).

Despite earning a reputation as "the best stockmen in the world," Aboriginal workers were "excluded from the provisions of industrial awards and worked and lived under conditions that would not have been tolerated by a white workforce" (Hess 1994, p. 65). This is reflected in the following quote from an Aboriginal resident in the Pilbara:

"All through the 1930s, 40s and 50s squatters still depended on our cheap labor. Our families lived in humpies and shacks around the station and worked as saddlers, horse breakers, cooks, stockmen, cleaners, child minders, gardeners, fencers, builders, windmill fixers, well diggers and at many other jobs: 'Most of the men and women who worked for the station are dead now. They made nothing. Only clothes, tobacco, cigarette paper, matches, food rations, stockman hat, stockman boots, that's all they used to get. Never made money'" (Tim Kerr cited in YAC 2011).

Even though some older Aboriginal people look back on pastoral work with nostalgia, the reality of their situation is encapsulated in the following comment:

"The pastoral industry of its time did give people a lot of room to live in the two cultures, to play a useful role in the pastoral industry from a wide viewpoint, economic viewpoint, and enough space to live a fairly traditional life. That was only ever going to be of a certain time. I have no sense at all that that could have just gone on for ever, it had to change at some time and the change was very brutal in the 1960s and I think careless of the interests of the Aboriginal people, but it had to come to an end" (Chaney, Interview 2011).

Chaney is referring here to the repercussions of the Pilbara pastoral worker's strike of 1946, the first strike ever coordinated by Aboriginal people in Australia (Hess 1994) which resulted in the strikers losing their jobs: "…with the equal wage case, people were simply ejected and there were a lot of very brutal things done around that time. For example Moola Bulla which was a large cattle station owned by the state, the state sold it and the Aboriginal people were promptly ejected and left on the fringes of Halls Creek and Fitzroy" (Chaney, Interview 2011). For further information on Moola Bulla see Neate (2013).

Prior to the provision of citizenship rights in 1967, most Aboriginal people continued to work in pastoralism or alluvial mining; many Aboriginal men and women would pan for gold and tin in return for flour, sugar, and tea (Holcombe 2005, p. 110). Most individual mining tenements (whether held by Aboriginal or settler Australians) were, however, taken over by the large mining companies particularly with the discovery of iron ore.

3.1.3   MINING IN THE PILBARA: FROM EXCLUSION TO ENGAGEMENT

Early Days of Mining: Impact on the Town of Roebourne

In reviewing the experience of Aboriginal people under colonial rule, Fred Chaney commented that "it is hard for us to imagine the extent to which we have brought destruction on a culture and the long term impact that has." Prior to the provision of citizenship rights Aboriginal people were not "counted" as people, they were not allowed to vote, their children were taken away, and many in the Pilbara lived in tin sheds on reserves well away from white view. Some understanding of how mining impacted on Aboriginal people at that time can be gained from looking at the experience of Aboriginal people living around the town of Roebourne. An understanding of historical context is important here, not just to explain the brutal impact of European colonization, but also the equally forceful disruption caused by mining in the 1960s to Aboriginal people's way of life. As Edmunds describes it, mining "introduced modernity in its most destructive forms" (2012b, p. 157).

The town of Roebourne was established in 1866. It had a secure water supply from the Harding River (*Ngurin* in local language) and was close to the harbor at Cossack; by the beginning of the 20th century, it was predominantly a "non-Aboriginal town, with solidly built church, bank, courthouse, post office, hospital, (and) three hotels" (Edmunds 2012b, p. 154). By the 1930s, an area known as the Old Reserve was designated for Aboriginal people to live across the Harding River outside the township. The Aboriginal people living on the Old Reserve (in shacks, tents, or under trees) were subject to a curfew until the mid-1960s and were excluded from town between 6:00pm to 6:00am; children from the Reserve were not allowed to attend the town school until 1961 (Olive 2007, p. 51; Edmunds 2012b, p. 155).

The federal government lifted its ban on the export of iron ore in 1961 and by the 1970s new towns had emerged close to Roebourne—Dampier, Wickham, and Karratha—as well as six new mining towns in the eastern Pilbara (Walker 2009, p. 212).

One Aboriginal man, interviewed for a book about the history of iron ore mining (sponsored by Rio Tinto), said:

"In the mid-1960s, iron ore development boomed in our tribal lands and new mining towns grew up overnight. Thousands of single men flooded in to build railways and towns for Hamersley Iron but Roebourne was not ready for the boom and couldn't cope with it. There were more construction workers living in Roebourne's caravan parks than in the town itself, and after work hundreds of them came to town to let off steam in the pub. Our community just fell apart …" (cited in McNair-Holland 2006, p. vii).

In those days, many company representatives were not interested in employing Aboriginal people on mine sites, in whatever capacity. Bob Beeton, who grew up in Roebourne, remembers that many Aboriginal people were treated unfairly, "… the sad thing about it was a lot of the Aboriginal people got pushed to one side because people seemed to think they couldn't stand the workload, be reliable all that stuff" (cited in McNair-Holland 2006, p. 4). By the 1970s, there were nine company towns in the Pilbara from which Aboriginal people were excluded and this only crystallized the marginalization of Aboriginal people (Holcombe 2004).

By the 1980s, Roebourne was transformed from being a "white" town to a predominantly Aboriginal settlement. The mine workers moved out, the administrative centre of the Shire moved to Karratha and "Roebourne was relegated to the status of backwater, with a largely Aboriginal population and a major alcohol problem" (Edmunds 2012b, p. 155). The local government then decided to shift the Aboriginal people from the Reserve on the edge of town to state housing in the centre of town:

"At least twenty five families from the Reserve told the government they didn't want to crowd into the state housing village. They wanted to live in smaller family groups out in tribal country. The government didn't listen. They just said, we've built you some houses and we want you to live there now, next to the cemetery. We never believed in living so close to where our people were buried.

Living arrangements on the old Reserve were organised around tribal and family groupings. In the new Village people were put anywhere, so that family groups were split up into houses all over the place. You know, leadership was very hard in the Old Reserve after drinking rights, but now it was broken down altogether. What was left of our discipline and respect system on the Reserve just went to pieces in the Village.

The teenagers who had grown up under the pressure of mining development and the early years of drinking rights, were confused and angry, and getting into more and more trouble with the police. Many teenagers lost their lives during those first ten years in the Village.

All the government promises of the benefits mining development would bring to everyone in the Pilbara never seemed to reach us, but we were given more police, a new prison and a dam" (YAC 2011).

Excluded from work, treated with indifference by government, and deeply affected by alcohol consumption, the Aboriginal people of Roebourne lived through a period of despair characterized by increased numbers of deaths in custody. In the 1990s, Aboriginal people, who constituted less than 3% of Western Australia's population, comprised one-third of the state's prisoners (Grabosky 1989).

In instances where Aboriginal people were employed—in towns such as Wiluna, for example, in the Northern Goldfields—there also appeared to be less understanding of the cultural obligations on Aboriginal peoples to attend funerals or return home if kin were sick or in trouble, etc. In the 1960s there was little acknowledgement that for Aboriginal people "culture frames a sense of identity which relates to being the First Peoples of the land. For Aboriginal children, families and communities, culture enhances a deep sense of belonging and involves a spiritual and emotional relationship to the land that is unique" (Bamblett et al. 2010, p. 100).

Without that cultural understanding, some employers misconstrued absenteeism as a lack of commitment to regular employment. A retired male engineer that we interviewed reflects on his past experiences employing Aboriginal people:

"… we also employed some people on a casual basis and they were supposed to come out and do a secondary part of this construction work and it was a damn nightmare. The biggest problem was never knowing whether you were going to have one, none or ten people at work. I think that that is something that Aboriginal communities themselves have got to address in the way that some of them are addressing drinking by making themselves dry communities. I don't have the answer, but I'm saying that this is a real problem, that there won't be a real share in the mining industry unless you can depend on people to generally be there when they're supposed to be there" (Retired Male Engineer 1, Interview 2012).

Early Days of Mining: Impact on Remote Communities in the Pilbara

In first half of the twentieth century, the only Aboriginal people still living on their "country" in the Pilbara were pastoral workers. The Mardu of the Western Desert were one of the last groups of Aboriginal people to come into contact with settler Australians. When Robert Tonkinson, an anthropologist, first met the Mardu people of the Western Desert in the early sixties, many of them had minimal contact with white people or "whitefellas" as they called them.

"There had been some pastoral activity around there and people would come in off the desert and gone and worked for pastoralists, the usual sort of first contact stuff. But once the camel breeding

depot was there, eventually the government said it would be a ration distribution depot and then it attracted more people over time. So Jigalong became an Aboriginal settlement through the fact that Aboriginal people were coming out from the fringes of the desert…" (Tonkinson, Interview 2011).

Gold was discovered on Mardu land in 1972, with mining operations beginning in 1975 (EPA 2002, p. 1). Newcrest mining established a residential mining town, known as Telfer, with basic services on site. Tonkinson describes a visit to Telfer in the early phase of mining:

"…I went out there (Telfer) in 1974 and we took some Aboriginal guys out there before when it was just a little base camp and spent some days going around seeing the sites and seeing the rock …when Telfer was originally set up, that was a closed town and Aboriginal people were absolutely forbidden to go there and if they did go anywhere near there they were certainly not going to get any petrol. Other (towns) have been similar in the past. Some companies have had a total conversion from totally antagonistic to anything Aboriginal to realising that there are political realities you've got to deal with eventually and you'd better change those views and over time they've changed" (Tonkinson, Interview 2011).

These kinds of closed mining systems in Western Australia were common when companies imagined they were operating according to the colonial doctrine of *terra nullius* (empty land). In such operations "everything even down to residency requirements for employees were part of the system of production, and we could control all of that. Even the Pilbara in its earliest days with company towns and closed roads and private rail loops and private ports, it could be controlled as a total system of production with no degrees of freedom beyond what we actually owned and operated ourselves. Well it's not like that anymore" (Hodge, Interview 2011).

Changing Attitudes of Mining Companies

The attitude of mining companies toward Aboriginal people at the start of the mining boom in the 1960s is evident in the accounts set out above. The mining companies were less concerned with "community engagement" (a phrase unheard of at that time) than with the speedy implementation of their operations: according to Rio Tinto's official website, for example, it took an extraordinarily short period of time—19 months after iron ore was first discovered—to commission the Mt. Tom Price mine infrastructure, building a shipping port in Dampier, a railway, and two towns (Capitalco n.d., p. 3). When describing how these operations were set up in the mid-1960s, one Hamersley Iron employee at Tom Price explained that "Hamersley had no problems with Aboriginal people because there were no Aboriginal organizations that had to be dealt with" (Edmunds 1989, p. 47). However, this perceived absence of Aboriginal organizations can be linked more generally to a lack of understanding about Aboriginal communities. As suggested by a Senior Aboriginal woman,

Verna Voss, in those days there was "no consideration for engaging the Aboriginal community and discussing with them whether it was appropriate to certain activities in certain areas because there was no regard for Aboriginal culture in that sense" (Voss, Interview 2011).

Beginning in the 1980s, the attitudes of mining companies toward Aboriginal communities in Western Australia began to change. This represented a wider shift in the resource sector, summarized by Ballard and Banks (2003, pp. 288–9): the explosion in mineral prospecting in the 1970s and 1980s often took place in remote areas populated by marginalized and/or Indigenous communities and this coincided with increasing acknowledgment of Indigenous issues and lobbying to acknowledge Indigenous interests in the resources industry (Langton, 2012). This was further bolstered by evidence of the "extensive loss of lives, livelihoods and environments" (Ballard and Banks 2003, p. 289) associated with mining (e.g., the Bougainville rebellion in PNG; the riverine dumping of the Ok Tedi mine; and human rights abuses associated with the Freeport McMoRan mine in West Papua). The widely publicized Rio Declaration and Agenda 21, held in Rio de Janeiro in 1992, also called for public disclosure on environmental impact (UNEP 1992). This initial focus on acknowledgement of physical impacts expanded to include social impacts, a move which has been partially attributed to the influence of international and national non-government organizations (NGOs) (Perez and Sanchez 2009; IISD 2013).

These developments led to a realization amongst Australian mining companies that it was no longer possible—or profitable—to not negotiate with Indigenous communities. Furthermore, for the first time, more than 90 years after federation, Australia's highest court recognized that Indigenous Australians had, and in some cases still have, legally recognizable rights to land. In the case of *Mabo v. Queensland* (No 2), decided in 1992, the High Court of Australia held that "the common law of this country recognises a form of native title which, in the cases where it has not been extinguished, reflects the entitlement of Indigenous inhabitants, in accordance with their laws and customs, to their traditional lands" (Howitt 1997).

However, the transition from exclusion to engagement with Aboriginal people was not smooth or easy. Fred Chaney recalls this as a difficult time:

> *"Well miners used to be the great adversaries of Aboriginal rights and interests. I mean I was Minister of Aboriginal Affairs then and there was considerable hostility and I'm talking thirty plus years ago, considerable hostility from the mining industry. But even the 'better end' were hostile to the notion that Aboriginal people had any rights that they needed, that they had a place at the negotiating table and really the good black fella was as far as they were concerned the blackfella who said 'yes boss'. I'd have to say my experience was of having to deal with those companies on a basis of considerable opposition to acknowledging the interests of Aboriginal people and the remarkable changes that were afoot"* (Chaney, Interview 2011).

The fearful reaction to granting Native Title in Western Australia is clear from the publication of news articles such as "WA Blacks Make Huge Land Claim" (Graham n.d.) and in a statement by the then Premier, Richard Court, who stated that if Native Title were not significantly extinguished, the nation would "become a laughing stock" (Hughes et. al. 1997).

Many mining companies saw Native Title as an expensive deterrent to exploration but Leon Davis, CEO of Rio Tinto from 1997–2000, believed it was necessary to embrace the reality of Aboriginal rights to land. He shifted Rio Tinto's public policy position from paternalistic "good neighbourism" toward a more critical and constructive engagement with unresolved grievances. His approach was unusual at a time when many other companies saw Native Title as an expensive obstruction to the process of mineral exploration (Howitt 1997). Many Rio employees at the time couldn't understand why the company wanted to change its policy, such as setting up an Aboriginal training unit as a preparation for working on mine sites. "They weren't hostile to it," one engineer commented, "but they were perplexed—why do we need all this stuff?" (Male FIFO Engineer, Interview 2011).

Janina Gawler, currently the General Manager (Communities) at Rio Tinto, attributes much of the cultural change within the company to the leadership of Leon Davis: "he gave a seminal speech to the Australian Securities Institute about the opportunities that would come if Rio Tinto, then known as Conzinc Rio Tinto Australia (CRA), as a company engaged differently with Aboriginal people. It was in the midst of the Native Title furore and the company stood aside from what had been pretty much an embattled ground of opposition to Native Title and seeking to find changes to the legislation rather than working with it" (Gawler, Interview 2011).

Gawler is unequivocal that this cultural shift within the company was built on a business case:

"… Rio's business case was that internationally we were working differently in places like Africa and had a track record of engaging with Indigenous people or people who owned the sovereign nation and indeed were the people of that country, but we seem to have a had a different view in Australia in terms of Aboriginal participation. What Leon did was basically to say if we can do it elsewhere why aren't we doing it in Australia?" (Gawler, Interview 2011).

The change in some mining company attitudes toward Aboriginal people coincided with changing attitudes amongst Aboriginal people themselves. Two Aboriginal activists and academics—Marcia Langton and Noel Pearson—typify a growing belief that Aboriginal people should no longer be considered as separate enclaves of "traditional" culture within the mainstream economy. Langton claims that most government policies toward Aboriginal people are "exceptionalist initiatives," which set these communities apart from Australian economic and social life (Langton 2012). She further argues that Aboriginal people should maximize opportunities from mining companies to chart their own future in remote Australia.

An educational program was implemented at Rio Tinto whereby managing directors and senior levels of leadership undertook cultural awareness training developed by Noel Pearson and Marcia Langton.[9] The initiative to employ Aboriginal people on mine sites then became a strong focus of company policy: "The fact that no locals were working with us who were Aboriginal people I think that was really big question mark. If people are not prepared to work with you, what is it about your culture in your organization?" (Gawler, Interview 2011). The response to this initiative is considered more fully in the sections below.

Since the implementation of Native Title in Australia, and the changing attitudes described above, it is now standard practice for mining companies to form agreements with Aboriginal communities. While agreements are unique to each Native Title representative body and region, the parties involved usually attempt to reach consensus on the following issues (ISS 2001, p. vii):

1. Employment and training;

2. Provision of education (this is primarily the responsibility of government, but agreements can play a role in improving outcomes through programs to encourage school attendance and for scholarship support for upper secondary and tertiary education);

3. Compensation arrangements (these are usually held in trust);

4. Equity arrangements in mining ventures and/or other joint business ventures; and

5. Protection of cultural heritage.

It should be emphasized that Aboriginal people theoretically have the option of trying to stop a project and not entering into negotiations, but this option is rarely exercised. As O'Faircheallaigh points out:

"… *Aboriginal communities are increasingly entering into negotiations with mining companies and with State agencies involved in the project approval process. It should be stressed that their decision to do so need not reflect either support for mining or an expectation that negotiations will yield substantial benefits. It may just as easily result from a desire to obtain some benefit from a project which Aboriginal people would prefer not to proceed with, but which they cannot stop; and/or a desire to avoid negative impacts associated with a project, particularly in terms of damage to the environment, cultural heritage or significant sites*" (O'Faircheallaigh 1995, p. 2).

[9] Professor Marcia Langton has openly advocated that Aboriginal people should benefit from Indigenous Land Use Agreements and from the employment opportunities offered by those agreements. She is employed at the University of Melbourne and has secured collaborative research projects with Rio Tinto (see http://www.atns.net.au/). Her views are not shared by all academics, NGOs, and other Aboriginal activists. See Frankel (2013) and Crook (2013) as an example of the debate generated by Langton's Boyer Lectures in 2012.

While there is a consensus that such agreements are necessary and that they "make good business sense," others link compensation to a wider sense of moral obligation and justice. The following excerpt from an interview with Fred Chaney, reveals a view that these Agreements are a form of redress for an activity that inevitably negatively impacts on Aboriginal economy and society:

"but how do you compensate people in a way which will actually wind up with them being better off, and that's why these negotiations are complex and where the miners in my view, the better miners let me stress, the better miners put a lot of emphasis on trying to ensure there are long term benefits. That the loss of value, of their native title, and of their connection to country is matched with a compensating fund of some sort, assets which will be available to their children and grandchildren and the people… and allying that with measures which help people live in a different world so education, employment, training and all of those things, business formation, all of those things really matter. But I mean these are matters of simple justice really. I mean, to underestimate the pain that all of this is causing people in the sense of their culture and beliefs is to ignore something that's very real" (Chaney, Interview 2011).

Measuring the Outcomes of New Initiatives

In 2010, Marcia Langton visited the Pilbara as part of her research into Aboriginal engagement with the mining industry. She noticed that Karratha had "new brick houses, tree-lined streets, substantial amenities, a motel, shopping centers, restaurants and tennis courts"; Roebourne, on the other hand, was "old, dusty, and showing signs of years of neglect: broken fences, potholes, weeds and flaking paint" (Langton 2010, p. 48). She also observed that both settler Australians and Aboriginal Australians who did not work for the mining sector in the Pilbara were at the mercy of high rents, lack of training, and generally insufficient opportunity to change their lives in a productive manner. Langton identified conditions in the Pilbara as emblematic of the "resource curse," a concept made famous by Auty (a geographer) to describe the phenomenon of resource-rich countries experiencing poor economic growth, increasing social conflict, and decreasing democratic standards (Auty 1993). While acknowledging that the "resource curse" is open to debate, Langton suggested that "the paradox of plenty" (2012, p. 23) typifies some Australian mining regions whereby neighboring Aboriginal populations face extreme poverty. Langton believes that it is possible to counter the effects of poorly distributed wealth in mineral-rich areas by making institutions and governments more accountable (Langton, 2012; Stiglitz, 2006). However, it seems clear to Taylor and Scambary (2005) that the wealth generated by the agreements is not translating into positive outcomes.

In the mid-1990s, less than half of one percent of Rio Tinto's Australian workforce was Indigenous but by the early 2000s Pilbara Iron had raised the level of Aboriginal employment to 8% while the Argyle diamond mine had increased Aboriginal participation to 25% of their workforce (Rio Tinto 2007, p. 6). In the case of Rio Tinto Iron Ore, it is fair to say these employment initiatives appear to have had mixed success. In partnership with the Centre for Aboriginal Economic Policy and Research at ANU, Rio Tinto commissioned research into the impact of its agreements, particularly in the area of employment. The report, based on sets of data about relative income status, participation in the workforce, mortality, and crime figures revealed that despite the creation of numerous agreements between companies and Aboriginal people, " … for a complex set of reasons, Indigenous economic status has changed little in recent decades—dependence on government remains high and the relative economic status of Indigenous people residing adjacent to major long-life mines is similar to that of Indigenous people elsewhere in regional and remote Australia" (Taylor and Scambary 2005, p. 1).

The Taylor and Scambary report concluded that the drive to employ Aboriginal people in the mining industry required tackling the "much deeper structural hurdles" of poor literacy and numeracy, low school participation, and lack of motivation (2005, p. 147). The authors also acknowledged that working in the mainstream economy was not the only option; working in land and sea management, the art industry, and cultural education were areas where many Aboriginal people preferred to work and some combined this with intermittent work on mine sites in conjunction with returning to country for ritual matters. These kinds of choices constitute, in Altman's terms, a form of "hybrid economy" which is more suited to those living in remote regions (Altman 2005).

3.1.4   ABORIGINAL RESPONSE TO MINING

In this section we look more closely at how Aboriginal people have responded to employment and business strategies in the Pilbara. Their views, along with others who work closely with Aboriginal people, reveal an ambivalence that explains at first hand why well-meaning initiatives to incorporate Aboriginal people have had mixed success.

Mining as Opportunity?

The Taylor and Scambary Report on the demographic profile of the Pilbara found that "despite substantial growth in economic activity and employment opportunity in the Pilbara since the 1960s … the overall employment rate for Indigenous people rose only slightly from 38% of all adults in 1971 to just 42% in 2001. This compares with equivalent figures for non-Indigenous adults in the Pilbara of 81% and 80% respectively" (2005, p. 27). It may be hard to imagine why the strategies which many Pilbara mining companies have put in place—education scholarships, training opportunities, work opportunities—are not grasped or embraced enthusiastically by Aboriginal people living in regional and remote areas. The history of Aboriginal people at Roebourne, described above, gives some sense of the depth of alienation experienced by Aboriginal people in northwestern Australia. Their continued marginalization is reflected in a well-documented disparity between Indigenous Australians and white Australians in terms of health and social and economic outcomes (AIHW 2013).

In 2001, for example, approximately 70% of non-Indigenous residents were in full-time secondary school education at age 16, compared to 44% of Indigenous residents (PHIDU 2005, p. 5). So while many Aboriginal people in the Pilbara, particularly in regional centers such as Karratha, Port Hedland, or Roebourne, would like to see their young people get local work, it is difficult to achieve if they have not finished Year 12 and are not motivated enough to join a training program.

> *"We are trying to get our kids into those mining training areas to get ready for jobs coming up. Some of them don't have the education, or too much drugs or alcohol. We try to counsel them and say you got to get into those jobs, you need to prove yourselves and stop all this other nonsense. Lot of people don't get jobs because they are right without, particularly with access to drugs and alcohol these days. They don't want to work. That's a big problem"* (Taylor and Scambary 2005, Interview Segment 1, p. 57).

There are many Western Australian Aboriginal people, especially elders, who are striving to change the attitudes of their young people (MacCallum et al. 2006; MacCallum et al. 2010). A large number of Aboriginal people want their communities to opt out of the "welfare" mentality and believe that working for mining companies could provide a solution:

> *"You find this anywhere, white or black. Easy street is what our younger generation prefer these days, everything easy, sit on the dole why work, why go and slave yourself around you see? But sometime we make 'em work for it though by going on those survey and heritage for mining with the anthropologists, all working, and that is a hard job! ... whereas if you put 'em here just cleaning the maid rooms, which I've been doing, I learn nothing from it, I'm just making an income to live"* (Taylor and Scambary 2005, Interview Segment 22).

There are many obstacles to changing attitudes toward Western institutions such as improving Aboriginal students' regular school attendance, with tertiary education as a long-term goal. Family and kinship ties often take precedence over whitefella obligations (Burbank 2006). One Aboriginal woman recalled the opposition to her wanting to travel away to Perth to study:

> *"I ended up marrying young then decided when the children came along I just wanted to do my teaching training so I went and did that. It wasn't easy, because family is so important with our people and keeping that closeness, and family feared that if I went away, I won't return and I broaden my horizons in that direction, it's going to break that bond in a sense. So I didn't have that support to go and do my training until I was able to speak to Dad and he saw that I was really serious about it and that I was serious that I would come back home"* (Voss, Interview 2011).

These types of cultural factors may also inhibit Aboriginal people from working on mine sites. Although common values, such as respect, may exist between an institutional culture of mining and Aboriginal people, there are common differences, such as systematic written systems on mine sites, with Aboriginal cultures often characterized as event-based and oral; whereas a mining culture may value an individual work ethic, an Aboriginal culture may feature collective work sharing (Gibson et al. 2011).

Reconciling Local Cultural Values with the Mining Economy

Some Aboriginal people, as well as anthropologists who were interviewed for this publication, expressed concern about Aboriginal peoples' ability to reconcile core cultural values with work on mine sites. The two concepts which we will address below are that of "homesickness" and "sharing."

Aboriginal groups were nomadic in pre-colonial times, with rights to hunt or gather food within a defined area, and more importantly they had a deep spiritual connection to that land. This unique connection between land, law, and culture was not legally recognized prior to the Mabo case (Hill 1995). Ties to country are fundamental to Aboriginal people and particularly so in the Pilbara, where some groups have been able to remain "on country," unlike the majority of Nyoongar people to the south.

In this context, being asked to work on other people's country can be seen as a deterrent rather than an opportunity. Indeed, "…a job or training program which requires movement away (physically or culturally) from a network of kin is a high-risk economic proposition for many Indigenous people" (Schwab, 2006, p. 15). Tonkinson explains the idea of homesickness:

> "…for the desert people, the pull of country to bring people back, it is still incredible because when you are on country you feel safe, you feel in command of stuff, being on country is a huge asset. Kids would jump out of dormitory windows at Port Hedland when they were going to high school up there and walk the 500 miles back to Jigalong and they would turn up back in Jigalong and people would understand perfectly. Nobody would say 'my God what are you doing back here?'… Being homesick is a sickness, a proper sickness, a real sickness and it's fixed by being back on country and so that is the thing that you can't see but is an enormous factor when let's say you are employing Aboriginal people out of their country for instance" (Tonkinson, Interview 2011).

There are also cultural transgressions associated with working on other people's country. Such feelings were also reported to us by a mine manager of an iron ore mine. He described in an interview how many new Aboriginal employees, who were not local to the area, could not sleep at night until a local elder was brought in as a mentor and to appease their feelings of "trespassing" on other people's land.

This same manager, a civil engineer by training, talked about some of the difficulties, challenges, and rewards of engaging with Aboriginal people. In order to meet his employment target, he employs Aboriginal people from the neighboring Kimberley region on a fly-in fly-out basis. A plane flies from Broome to the mine site every week. When the company wanted to change the rostered work arrangements so that the men would have more time off in between work commitments, the manager told us that the elders at One Arm Point would not agree to it. "Most of these elders are ex-stockmen…they know the meaning of hard work. They believe that if young men have too much time off, they have more time to get into trouble" (Mine Manager, Interview 2011). He acknowledges that Aboriginal elders are troubled and conflicted by FIFO work arrangements: on the one hand, it takes young people away from their community and they are less likely to learn about and maintain their culture; on the other hand, work brings some idea of dignity. This particular mine manager believes this tension is manageable as long as we "work on a small scale" and build up trust with the community.

Entrepreneurship and Sharing: Different Views

Sharing in Aboriginal communities forms part of a complex cultural system linked to reciprocity; this type of sharing can provide economic assistance between Aboriginal groups and shape or deny social networks (Schwab, 2006). How do Aboriginal ideas about sharing and kinship obligations fit within a conventional Western economy in which entrepreneurship is often associated with self-interest, and decision making is ostensibly guided by the need to make profit? Tonkinson claims that "today, the arena of Aboriginal affairs seems to be an ideological battlefield that pits neo-liberalism's obsession with the individual agent against a deep-seated emphasis on the group in Mardu and many other Aboriginal societies" (2007a, p. 42).

One aspect of this "emphasis on the group" is expressed in ideas about sharing. In conversation Tonkinson says that

> *"the way money circulates in hunter-gatherer societies, the share thing, it circulates like crazy. It's all about kinship, it's all about obligations and a responsibility to other people and claims that you have on them and they have on you. So that doesn't go with our entrepreneurial spirit. Our entrepreneurial spirit is that you are this person who is free to develop your skills and to make yourself, you know, the Horatio Alger[10] story which Americans still believe is possible for everybody"* (Tonkinson, Interview 2011).

These views about sharing amongst Aboriginal people were reinforced by other people in the interview group. Don Miller is a former pastoralist and an Anglican pastor, who runs a station at Ulalla on the border of the Pilbara and the Northern Goldfields. He has had a long relationship with the local Aboriginal people who now live on his property and propagate seeds for use in the rehabilitation of mine sites. Don commented that possessions were "destructive" for Aboriginal people, an idea that has been put forward by Victoria Burbank in her writing about life in a remote settlement in the Northern territory. After thirty years' fieldwork in one settlement

---

[10] Horatio Alger, Jr. (1834–99) wrote many "rags to riches" stories in the 1860s, usually about street boys living in poverty who overcome adversity to achieve wealth and status (Decker 1997).

(Numbulwar), Burbank (2011) observed that there was deep tension between the emphasis on the group in Aboriginal social life and ideas about individual ownership, particularly ownership of "whitefella" goods.

Verna Voss has a slightly more nuanced view about how Aboriginal culture operates. Commenting on Bob Tonkinson's view (above) she says: "I see where Bob's coming from because within our traditional cultural society it is very much a collectivist society; individualism is not part of the culture. So that's where Bob's focus is from that anthropological thinking and he would be thinking of a lot of the older people he worked with out there at Jigalong" (Voss, Interview 2011). However, Voss suggests there is a great deal of diversity in Aboriginal views on how to "move forward" in engaging with "whitefella" ways. She acknowledges that many Aboriginal people find it difficult to get around "the sharing and obligation which are key elements of our [Aboriginal] culture." Equally, Voss states that there are others who "may have been more exposed to Western structures through the schooling system and things like that, so that cognitive development in that area and analysing things in as different way will happen sooner that they start going into the enterprises and those sorts of positions" (Voss, Interview 2011). Voss feels that it is important for companies to understand the cultural diversity within Aboriginal groups. While her brothers combined work on pastoral stations with work on mine sites, there are many other Aboriginal people who would choose not to work on mine sites:

> "… that's why companies need to be mindful that there are all different levels and there are many that don't even want to go into a mine to work because that in itself is in conflict with the belief and the attachment to land and what it's going to do to the land" (Voss, Interview 2011).

Voss also believes there are practical ways to manage the obligation to share, particularly in relation to cash wages:

> "You can have two bank accounts. You know you're going to always share something but work out how much you're prepared to give as a gift. You're never going to get it back you're going to see it as a gift. Put that in that account and if you're saving for a car or for your bills and that you put that in this account. The same thing I said 'if you're not working, you know you're still wanting to share; sometimes when specials are just get some extra flour and different things and have them in the cupboard. You don't have to give money, but then if you know that person's got kids and they haven't got much to feed them and you're worried about that and that's what's keeping you back, then you can give that extra food' and things like that. Traditionally it was reciprocal. Sometimes now it's not, it's a one-way street because people are thinking 'well if you're working, you've got more to give than someone else'" (Voss, Interview 2011).

There are other Aboriginal people who contest the view that Aboriginal culture cannot accommodate Western economic practices. Barry Taylor was the CEO of one of the largest Aboriginal owned and operated contracting companies in Western Australia (Ngarda 2012). He is an outgoing and entrepreneurial Nyamal man from the Pilbara region who recounts that "all his life has been impacted by mining" (Barry Taylor, Interview 2011). His father is a well-known character who led the first Aboriginal strike for better wages on a pastoral station in Australia; when he and others walked off the station, they took to panning for tin and gold. Barry suggests there was resistance (by Aboriginal people) to enter the mining industry in the 60s, 70s, and 80s but believes that generational change has occurred.

With a vision encompassing Aboriginal employment by mining companies and the development of Aboriginal owned and operated mines, Taylor strongly believes that Aboriginal people have to "get off welfare" and his goal encompasses, not just employment by mining companies, but the development of Aboriginal owned and operated mines. He argues that "mining is the way out" for his people. "We have to save our people," he says and "welfare is no good, and cottage industries don't work" (Barry Taylor, Interview 2011). He also sees no contradiction between being a "law man" (that is, practicing core Aboriginal values) and being an entrepreneur.

A similar view is expressed by Fred Chaney:

"being a true entrepreneur is a rare thing in any society. Most of us are not entrepreneurial, you're not. I'm not. Small business people are people who are driven by a set of values and an approach to life which is not given to many. Most of us are wage slaves or conduct a very small business through our own efforts and so on. I think to expect Aboriginal people with such a different background and an educational background, in many cases a lack of education, maybe in our sense to be entrepreneurial is silly" (Chaney, Interview 2011).

A similar belief—that it is not an innate skill to be entrepreneurial—is expressed by Don Miller who feels that it is "stupid to try and involve Indigenous people as miners" and goes on to explain why:

"...there are always going to be the occasional person that it's going to work for no matter what and that doesn't relate to Indigenous people, it relates to anybody. There's always going to be the different person that's going to take up that thing that nobody else wants to do. So there's that, but to say 'this a way forward in life' is crazy because even for whitefellas we're really only doing it to an end. Somebody I know really well is a mine engineer who is now a mine manager and way up at the top of it and he would still wander around with a T-shirt saying 'I was born to fish' and that really is where life is and this is only a means to somehow get to that. The majority of Indigenous people wouldn't see that they need to be miners to get what they want to be. It's as simple as that" (Miller, Interview 2012).

This debate touches on some very profound issues about the relationship between culture and economic behavior[11] which will not be elaborated on here. For the purpose of this publication however the main point is: given that mining and resource development are considered inevitable and long term, it is important to remember that not only are Aboriginal ideas about work, entrepreneurship, and distribution of wealth different from Western ideas but neither are their meanings shared within Aboriginal societies.

Damage to Land, Damage to Culture: Concern about Water Usage

Mining operations account for an estimated 26% of total water use in the Pilbara (Department of Water 2010, p. 7). These figures, in conjunction with predictions that water availability from the region's surface water and groundwater sources is likely to become more variable in the future due to a drying climate trend (Department of Water 2010), are therefore of concern to the state government and mining corporations. Aboriginal people are no less concerned about water usage, but for very different reasons.

Water is clearly necessary for survival in semi-arid regions such as the Pilbara. But for Aboriginal people, it has more than material value: "water emphasises the interconnectedness of places from an Aboriginal point of view and associates the material and the economic with notions of sociality, sacredness, identity and the giving of life" (Jackson 2005, 2006). In many accounts from across Australia (Barber and Jackson, 2012, p. 33, Rumley and Barber 2004), water's vitality for unifying the mythical and material to sustain Aboriginal lives is evident.

In their 2011 report for Rio Tinto, Barber and Jackson (an anthropologist and an environmental scientist) wrote about Indigenous perceptions of water in the Pilbara. The following comments from Aboriginal people in the Pilbara about the significance of water reflect fears, not just about the loss of water, but also about failing in their role as cultural custodians:

> *"Water is life. It is sacred, certain pools, guardians, water serpents, in waterholes and permanent springs. The rivers all contain significant sites and there are serpents in the water. The elders are telling stories about the water needing to survive. Letting things like big developments happen, it's like us not looking after country. We're not scientists but we can see what is happening"* (Brendan Cook, cited in Barber and Jackson 2012, p. 27).

> *"Water provides fish, plants, wildlife, hunting and recreation. It's just drinking and washing for the white man but for us it provides everything. I remember we used to have Christmas*

[11] This issue was raised in the early part of the 20th century with *The Protestant Ethic and the Spirit of Capitalism* in which Max Weber identified the rational organization of labor, and the rational calculation of investment for long-term profit as the unique feature of Western capitalism. The idea that an entrepreneurial spirit is innate to European, North American, and Australian societies has however been challenged by thinkers such as David Graeber who argues that the assumption that we are all "maximising" individuals is contradicted by the co-existence of gift economies and a spirit of altruism that exists within capitalist societies (Graeber 2001).

in the Fortescue area. Just drinking water, swimming and fishing. Then there is the mythology side. The Rainbow Serpent is the giver of life, it looks after the waterholes. If the serpent goes, then the water goes, and then life goes" (Cyril Lockyer, cited in Barber and Jackson 2012, p. 28).

The appearance of sinkholes has been associated with extensive dewatering in other parts of the world (de Bruyn and Bell 2001). Although this has only been reported anecdotally in the Pilbara's Karijini National Park (Keane 2010), the following comments were made to Barber and Jackson:

"There are huge sinkholes opening up on Millstream country and in Karijini. The older people say that this is the snake, who is not happy and who is looking for his water" (spokesperson for the Ngarluma Aboriginal Corporation, cited in Barber and Jackson 2012, p. 23).

"When they take too much water, then the serpent gets upset and leaves. Once the rainbow serpent goes, then the water goes with him. And we all miss out. And Aboriginal people are the ones who get punished. We are supposed to be looking after him. They can take water, as long as he's satisfied that he's not being dried out. They have to do the monitoring so that everything else survives. They have to have limits on the extraction licences" (Cyril Lockyer, cited in Barber and Jackson 2012, p. 23).

What we see here are two different groups expressing concern about water usage in the Pilbara. For governments and mining corporations, it is essential to conserve the existing water resources. This can be done through legislation (water restrictions) or engineering innovations such as recycling water on mine sites, or constructing desalination plants on mine sites to avoid wastage in the dewatering process. These strategies serve to ensure economic sustainability. Aboriginal views about water, however, are based on a worldview in which culture, kinship, and landscape are interconnected; for Aboriginal people, the conservation and care of water is about cultural life.

Final Comments

The engagement between mining and Aboriginal people in northwest Australia began at a time when Aboriginal people already suffered great disadvantage. Unlike the people of Papua New Guinea, who became sovereign rulers of their own nation, the Indigenous people of the Pilbara had barely been granted citizenship rights when the iron ore "boom" began. Over the years, mining has presented Aboriginal people with difficult decisions: "It can have devastating effects on the environment, and negative cultural and social effects can be associated with its impact on sites of cultural or religious significance, and with the in-migration of mineworkers. But mining can also generate employment and business development opportunities and substantial revenues for Indigenous peoples, assisting them to overcome social and economic disadvantage and offering an economic base on which to maintain their cultural and social vitality" (O'Faircheallaigh 2012, p. 4).

Thus for many Indigenous peoples, the resource industry is a complex terrain that must be carefully navigated in the hope that its positive aspects can be maximized and its negative impacts minimized. Not all Indigenous people have a common view of how this balance can or should be achieved (O'Faircheallaigh 1995).

3.1.5   ABORIGINAL IMPACT ON MINING AND MINERS

It is more common to imagine the social impact of mining occurring "on" or "to" local communities, whether Indigenous or non-Indigenous. Although this an area which deserves further research, we would like to point out there is some indication that non-Aboriginal people who work on mine sites—including engineers and non-engineers—can also be affected by exposure to Aboriginal belief systems. The opportunity for social interaction was—and possibly still is—limited. The system of shift work within the framework of fly in/fly out or drive in/drive out can be a deterrent to white employees coming into contact with other Aboriginal employees in particular or with local communities generally:

> *"Working life like this was 14 days, 12-hour days, 6.00 in the morning till 6.00 at night. You talk for a while after work and then you go to bed or, if you need to, you go back to the office and work. What I'm saying is, that it's a terribly constricted sort of existence. You really only have to do with the people that you're working with, which—there's technical problems and they're really interesting, but geologists often got the opportunity to wander around a bit more, surveyors too, but engineers tend to be locked up in an office with a computer and told to get on with it"* (Retired male engineer 1, Interview 2011).

Before the introduction of cultural awareness training, discussed below, few employees learned about Aboriginal culture and society, and the opportunities to do so came up randomly, without planning:

> *"When I was on the environmental side, I actually got involved with the local people because of the sacred sites. I went out on one occasion with an anthropologist and another guy and we went out with about five local elders, looking at various sites that we wanted to mark on our plans that if the geologists went near them they had to contact us first. Some of it was strictly taboo for females. But really going around this area and then pointing out sites of where all these mythical sites and what happened here and there at different stages, it was just – I hate to use the term mind-blowing, but it was just so interesting those couple of areas"* (Retired male engineer 2, Interview 2011).

It is difficult to make claims about the impact of cultural awareness of training on those who work in the mining industry as no research has yet been undertaken in this area. However, it was clear from our interviews that individual response to this kind of training is variable. Some, like the engineer above, find it interesting; others are not affected by the training at all. Verna Voss has adapted her material to different groups but will generally cover the kind of material which this chapter has presented: the impact of colonization on Aboriginal people, Aboriginal ideas about land, and Aboriginal ideas about sharing and kinship:

> "*I read them when they walk through the door and I think, yep, he's come kicking and scream-ing, that one there really wants to be here and whatever so then I've already profiled my group. Many of them will come up afterwards and say 'You know what, I really didn't want to be here, I thought it was going to be a waste of time but boy was I wrong and thank goodness, you've opened my eyes.' What does sadden me to hear sometimes with a lot of them saying 'we are ashamed to be white' when they hear some of the stuff and I say 'well you need to change that thinking. That is not going to be helpful. You've got to use this in a constructive way and that can only start with you looking at that and not going down that road and give yourself time to digest a lot of information'*" (Retired male engineer 2, Interview 2011).

In her opinion, this type of training can be "tokenistic" or it can reflect a genuine desire to engage with Aboriginal people and to address their local needs.

3.2   BODDINGTON, SOUTHWESTERN AUSTRALIA

The town of Boddington is situated 152 kilometres southeast of Perth. European settlers first came to the area in 1862 and the town was gazetted in 1912 (Shire of Boddington 2004). The main activity of the region is sheep farming. Gold mining in Boddington began in 1987, much later than in the Pilbara, and operations were extended after a new ore body was discovered in 1994; the "re-vamped" mine began operating in 2009 (ECS 2008, p. 7). Although Newmont Mining have entered into an agreement with local Aboriginal people, the Gnaala Karla Booja (Newmont 2013), we will focus solely on the impact on the non-Aboriginal community in the township.

3.2.1   RESEARCH MATERIAL

There is very little historical material available for this region and no ethnographic work has been carried out, either in the township or on the mine site. The *Studies in WA History* journal (founded in 1977) does not, for example, contain any articles that feature, or even mention, Boddington. With regard to the gold mine, which was first established in 1987, there is an early report about the impact on flora and fauna (Worsley Alumina 1999); various Environmental Impact Statements (for example EPA 1985, 1994, 2001) but no social impact studies were carried out until 2006 after Newmont took over the mine (Banarra 2012, p. 3). This relatively recent expansion of company interest from environmental to social impacts follows the global shift in corporate interests referred to above (see Section "Changing Attitudes of Mining Companies").

The material in this chapter is therefore derived mainly from interviews with residents supplemented with Shire reports. The aim was to use a purposeful sample to gain an in-depth understanding of a limited number of participants' views. The interviews were conducted with the following individuals:

- Owner of petrol station

- Volunteer on community newspaper

- Previous owner of pub

- Worker at the Worsley Alumina mine

- Worker at the Boddington gold mine (driver)

- Real estate professional

- Secondary school student

- Retirees from nearby town of Quindanning

- Retiree from Boddington

The interviews were anonymous, open-ended, and centred around the question "how has the gold mine impacted on you, your family or your community?" Some people declined to be interviewed but most were very open and forthcoming about their views.

3.2.2   EXPANSION OF BODDINGTON GOLD MINE IN 2009

When it was known that the gold mine would be re-opened, in 2009, the Shire of Boddington (henceforth "the Shire") conducted a needs analysis of how the town would accommodate an influx of mine workers. Written by private consultants, the report stated that

> *"Boddington Gold Mine hopes to minimise the resident camp workforce as quickly as possible and will encourage employees to live within 50 km of Boddington town. The opportunity to do this will be strongest in the next few years. After that, families will have made long term arrangements and will be less keen to relocate, although natural workforce turnover will see some moves into the area"* (ECS 2008, p. 1).

The report also reiterated the company's commitment to promote local employment and residence by stating: "The owners have undertaken to encourage local employment and residence rather than support a long-distance commute project as is the norm in much of regional Western Australia" (ECS 2008, p. 3). Newmont also promised to implement a "buy local" policy (Esteves et al. 2010).

In addition to the promises made by Newmont, the West Australian state government had previously argued for an economically robust regional economy, based on sustainable use of resources and partnerships between industry, regional communities, and government:

"Western Australian regional communities will be healthy, safe and enjoyable places to live, offering expanded educational and employment opportunities for their residents and a high standard of services. They will be healthy, safe, and enjoyable places to live and work, offering expanded and improved educational and employment opportunities for their residents and a high standard of services. Regions will have robust, vibrant economies based on the sustainable use of economic, social and environmental resources and a strong partnership approach within and between regional communities, industry and Government" (Department of Local Government and Regional Development, 2003, p. 10).

This vision of a robust regional economy was extended with the designation of Boddington as one of nine "supertowns" in the state of Western Australia. The creation of regional supertowns is funded by the Royalties for Regions scheme whereby 25% of the State's mining and onshore petroleum royalties, equivalent to approximately 5% of the State's budget, will be returned to the State's regional areas each year as an additional investment in projects, infrastructure, and community services (Department of Regional Development & Lands, 2011, p. 4).

Figure 3.3: The main street of Boddington (photo courtesy of E. Feinblatt).

Yet when one enters the township of Boddington, it is hard to believe that there is a bauxite mine at one end of town and a gold mine at the other (see photo above). The resident population of the township has grown from 1,379 in 2006 to 2,226 in 2011 (ABS 2013a). Yet there is one small supermarket, a small health clinic, one bank, and a bakery. There are no cafes, few retail shops, not even a butcher. Although there are monthly Sunday markets, similar to many small regional towns in Australia, even the petrol station closes after 12pm on a Saturday. So the question which begs asking is: how has large-scale mining impacted on Boddington's social and economic development?

3.2.3   COMMUNITY RESPONSE TO MINING

During several open-ended interviews conducted in Boddington, a number of residents suggested that the local community was muted in its perception of the gold mine outside their town; there was neither whole-hearted endorsement, nor was there outright antagonism. There was acknowledgement that the mine has provided work for locals at a time when it was difficult for rural people to find employment and this recognition aligns with the Shire's statistics illustrating a low unemployment rate (1.7% in 2011), compared to 4.7% across Western Australia (ABS 2011).

Figure 3.4: The Newmont Community Information Centre, on the town's main street (photo courtesy of E. Feinblatt).

Still, for some local residents, there appears to be a great deal of ambivalence about the long-term benefit of mining to the community. Compared to the experience of Aboriginal people in northern WA, the community of Boddington appeared to have a relatively strong and cohesive community life, facing challenges common to small rural towns: for example, when wool prices dropped and crop returns were variable (Haslam-McKenzie, 2009), many young people moved away from Boddington. Since Newmont's re-opening of the gold mine in 2009, some town residents feel that the mining company is distant and removed from community life and are unsure how mining can assist their long-term goal, which is to sustain a viable community in a rural setting. These views permeate the following themes which emerged in a selection of interviews with local residents.

The Separation of the Mining Enclave and Community Life

Generic terms such as drive in drive out (DIDO) or fly in fly out (FIFO) are used to describe

> *"...work arrangements for resource operations located at a distance from other existing communities. The work involves a roster system in which employees spend a certain number of days working on site, after which they return to their home communities for a specified rest period. Typically the employer organizes and pays for transportation to and from the worksite and for worker accommodations and other services at or near the worksite. While most operations fly their workforces to and from their worksites, other modes of transport may be used"* (Storey 2010, p. 1161).

Many residents oppose the DIDO model, whereby workers live in the host community on a camp site close to the mine and drive to and from their home communities. In Boddington, the opposition to DIDO appears related to numerous factors, such as mine workers not utilizing local services, there being little or no interaction between the mine workers and local residents, and the lack of incentive for mine workers to live locally.

As expressed in the quotes from interviews below, these factors relate to the economic and social tensions of having large work camps that physically separate the mine workers and the local Boddington community. More broadly, such tensions are linked to the high demand for suitable accommodation to house a new resource's labor force (Haslam-McKenzie, 2009) in a community which was historically dependent on agriculture and timber:

> *"They built their construction camp out there which houses 2,000 men which is more than the population of Boddington, and there doesn't seem to be a great deal of interaction or interest, probably the pub is the greatest interaction and interest"* (Volunteer at local community newspaper, Interview 2012).

> *"They'd worked out there for ten years and had never been to town because they don't have to come to town. They have everything out there. They have a wet mess, they have gyms, they get fed, they get looked after, their rooms are cleaned, the whole lot. It's like a hotel out there really, so the guys come down and their seven shifts and go home" (Owner of local garage, Interview 2012).*

"Everyone that lives on the camp, there's no (incentive) for families to come here. People can just come here and live in the camp for a week and go home, seven days, seven off. Why bring the family up to Boddington, an hour and a half from Perth. There's no incentive to stay here and it's just cheap for them. I always thought that mining camp was going to be shut down till to about a 500 man camp after construction, but it didn't work. There's still a massive camp out there and it's just changed the whole town. The towns gone backwards since I've been here seven years "(Single male worker, Bauxite mine, Interview 2012).

In Boddington, more than one local resident unfavorably compared the current DIDO system with previous employment at the bauxite mine, at the other end of town. In contrast to the gold mine, the bauxite mine is relatively small and has been operating since 1979. This mine previously employed mainly local people:

"The bauxite mine was different too, very much so, and probably 70 or 80 percent of the employees at the bauxite mine would have been the locals at that stage. That was, I'm not good with dates, 20 odd years ago, it was a long time ago and it wasn't a good time financially and that's where it was so good for the town because it then employed everybody, as I say including the shearers" (Retiree who has lived in Boddington for 20 years, Interview 2012).

"I've been here for 57 years. I came here as a school teacher, I married a farmer. We eventually came off a farm, built a house in town. I've seen lots of enterprises come and go. The bauxite mine was probably one of the best things that happened to Boddington. We were in a recession at the time and it absorbed most of the local population at the time. It also unfortunately absorbed all our shearers so we were left with a dearth of shearers. The gold mine, I don't really think as far as I'm concerned it's had a great impact" (Volunteer at local community newspaper, Interview 2011).

Boddington, which was historically dependent on agriculture and timber, is now home to two mining operations, with one of them—the gold mine—being a large-scale operation involving mobile workers. In this context, future challenges for the community consist of gaining economic benefits associated with mining while maintaining the town's social cohesion and sustainability (Petrova and Marinova 2013).

Working on a Mine Site: Transiency

Many people who were interviewed acknowledged that Newmont was responsible for increased employment and they were also appreciative of the scheme to train and hire local men and women to work between regular school hours (9am–2pm). Nonetheless, some local residents raised concerns about other aspects of mining activity, which may erode community life in Boddington.

Firstly, according to one long-term local female resident, even when some mine workers choose to live in town, they will not necessarily live there for a long time.

> *"The mining industry seems to run in three year cycles. You'll have a job here for three years and then you're going to move on somewhere bigger, better up the ladder. Different priorities with your family, maybe you get a house supplied and you can be with your family somewhere else. So it's very hard then to keep a stable population because it just doesn't happen. From the mine managers down"* (Volunteer at local community newspaper and mine worker, Interview 2012).

Secondly, a local business owner mentioned that there appears to be a constant movement of DIDO workers who stop briefly in town to buy petrol or have a drink in the pub on their way to and from shifts. None of these workers live in town, however, and do not have time to form friendships with local people:

> *"I guess for us as a business, it became super hectic. The hours were longer. There were people everywhere. They weren't locals, they were just bussing people backwards and forwards. We had a huge amount of people coming to town. I guess we were warned that it would change what happened in town for a start in that there was so many people with contractors. It wasn't like that country town had been where everybody knew everybody"* (Owner of local garage, Interview 2012).

A local real estate agent also commented that many DIDO workers did not appear interested in forging friendships with local residents:

> *"Very few, if any, friendships are formed between DIDO workers and local people …because they're so transient … They have no alliance to Boddington at all. They don't care what happens to it and their workers they're just as easy to just say right that's it I'm finishing. They've got no ties here and they go. So they don't have any alliance to the town whatsoever"* (Real estate professional, Interview 2012).

These concerns illustrate the complexities of having a large heterogenous mobile population encroach on a relatively small and homogeneous *local* population (Haslam-McKenzie, 2009). While some residents in Boddington may welcome the economic benefits due to increased patronage of local services, others appear concerned by a lack of attachment to the town on the part of the mobile workers.

Finally, whether they are resident or mobile, all mine workers are generally involved in long work shifts. Some may find shift work arduous, others may not; there are similar differences in whether a high income will compensate for disruptions to family life. Such differences in opinion are evident in Boddington and only one resident, for example, perceived shift work to be disruptive to community and sports life in Boddington:

"Unfortunately as far as the town goes, the sport goes, the 12 hour shifts have played absolute havoc with football teams and bowling teams and tennis teams, all this sort of thing because you cannot get the continuity of your players. They're off every third week, they're here and then they're not here, they're working" (Volunteer at local newspaper, Interview 2011).

To attempt to address such social and economic tensions in regional communities, the Standing Committee on Regional Australia (2013), which reported on the use of FIFO and DIDO work practices recommended that:

- The national Housing Supply Council urgently create and implement a strategy to address the issue of affordable housing in host communities;

- The Australian Small Business Commissioner find ways of enhancing the capacity of small businesses in host communities to assist in servicing the resource sector.

Strategies and policies exist but, to date, these do not appear to have been effective in maintaining the soul of a community when DIDO and FIFO become disproportionally large compared with the population.

Engagement with the Community

Newmont describes its guiding values as social responsibility, sustainability, and safety, with the aim to be "the Miner of choice for all stakeholders including our people (both employees and contractors), the communities in which we operate and our shareholders" (Newmont 2013). Nonetheless, when respondents were asked about the mining company's engagement with the community, several residents appeared dissatisfied with the way the company interacted with the local community. This viewpoint emerged across some interviews regardless of whether the respondent felt that mining was a positive or a negative change in their lives.

For example, one secondary student would like more local facilities that cater to youth's interests:

"There's not much for our age-group, sixteen to nineteen years old, to do. As soon as we get our P's we go off to Perth or Mandurah, sort of a bad thing.[12] The mine did open up a child care centre so if they can do things like that I'm pretty sure they can do other things. Recently we just built a BMX track down in the Park and I had a look at that and it was tiny. It was absolutely tiny and I was like yeah we're going to have fun on that. We could walk that and do bigger jobs than what a bike could do, but we're trying. I guess the towns trying and that's all that matters ... they need more involvement with the kids and not keeping us in the dark. We'd be much more appreciative of what they do. Knowing that they care and want us to be involved, not just work wise but community wise" (Secondary school student, Interview 2012).

[12] P's: probationary license plates for new drivers in Western Australia.

While acknowledging the contributions of the Shire and the mining company to improving local services, such as the child care centre and BMX track, this student would welcome additional overtures from the company, especially toward improving the lives of Boddington's youth.

In addition, a retired couple who are keen horse-riders wanted Newmont to open up part of their land to allow access for bush walkers and horse riding:

"The other thing we find as horse riders is that we're being constantly locked out of areas that we'd like to ride and it would be good to have not just riding trails but cycle trails and walking trails through some of the beautiful bush areas but they're just completely locked out to us so that is a negative from our point of view" (Retiree 1 from Quindanning, Interview 2012).

"We have written to the mine and actually asked them to consider allocating or allowing us to have a section of their fire break or the edge of their property running from the mine in Boddington right across to Albany Highway where it joins the bushlands there. If the shire could do that as a walk trail for people it would be there forever and they could actually hang their name on it if they wanted to, it would be a good PR thing" (Retiree 2 from Quindanning, Interview 2012).

These comments highlight the tensions between local communities and mining companies, featuring topics such as land use, safety, economics, and recreation. In the case of Boddington, although Newmont has initiated a community reference group, the gold mine is adjacent to the town and requires a major buffer to alleviate the risk of environmental mishaps (Haslam-McKenzie, 2009).

The same retirees highlighted a certain amount of scepticism about the idea of "community engagement" on the part of the mining company. Each retiree, for example, wondered if the company's consultation process illustrated genuine concern for the local community:

"I think they've got their own agenda. They do sort of have these consultations but I think it's, well that's my feeling it's just superficial, that they really know what they want anyway. That's my feeling" (Retiree 1 from Quindanning, Interview 2012).

"It's a PR exercise for them, they must do it but I don't know how much notice they really do take of community opinion" (Retiree 2 from Quindanning, Interview 2012).

The themes which emerged in the interviews from the Boddington sample reflect similar views in an earlier Social Impact Assessment carried out by a private consulting company for Newmont Boddington Gold. This report identified "broken promises" as a common phrase which reflected the community's disappointment that Boddington had not become the thriving social hub as envisaged when the mine re-opened in 2009 (Banarra 2012). Our interviewees acknowledged that Newmont had brought work to their town but also revealed some disquiet at the lack of real "engagement"—that is, that their concerns were not being heard.

3.3   DEBATES ABOUT MINING IN WESTERN AUSTRALIA

3.3.1   MINING AS PROGRESS

Despite the ambivalent responses considered above, there are many—who work within the industry, and who supply the goods and services associated with it—for whom mining represents "progress" and "opportunity." The Perth skyline is dominated by buildings occupied by companies such as BHP Billiton, Woodside, and Rio Tinto as well as the engineering companies which provide services to them. Iron ore, petroleum, and gold account for 89% of all mineral sales in the state, valued at $95.1 billion (DMP 2011). It is difficult to comprehend these types of figures, but the scale is in keeping with the physical magnitude of the operations themselves as well as the language of boom and bust used to describe mining ventures. This news report about the mining boom in 2010 was written in glowing terms:

> "As cranes swing across Perth's skyline, busily erecting towers of steel and glass, Australia's most isolated state capital is reinventing itself as the gleaming face of a mining boom. Static for a generation, new high-rises are now rapidly appearing across the cityscape, while the streets have had a cosmopolitan injection of hip shops, bars and restaurants, all built on rampant Asian demand for iron ore. Today, 20 per cent of office space is leased by four companies: mining giants BHP Billiton and Rio Tinto, and oil and gas producers Chevron and Woodside Petroleum, all major players in the vast state's resources industry" (Pascoe 2010).

The professionals who work in this industry such as geologists, hydrogeologists, geophysicists, and engineers are paid well. So too are the semi-skilled workers on the mine sites in the Pilbara and Boddington, including personnel such as haul-truck drivers, dynamite crews, and drillers. Most of these mining personnel are FIFO or DIDO. Working patterns include 8:6; 4:3 (presented in the form of days on site: days off) or 2:1 (weeks on site: week off). Work shifts are typically 10–12 hours per day with the majority of mine employees living in Perth, with some residing in regional centers, the eastern states or, even Bali.

3.3.2   LIVING IN A "TWO STROKE" ECONOMY

For many residents of Perth, there are negative consequences for living in what some commentators have called a "two stroke" economy; that is, a mining economy in which employees earn very high wages which may increase if demand for iron ore or gold continues, against the mainstream economy in which wages are considerably lower and do not increase without union negotiations or some other form of bargaining. Furthermore, the mainstream economy is affected by job losses as manufacturing industries close, or as tourism declines because of the (currently) strong Australian dollar. It is for these reasons that some commentators talk about the "boom and gloom" rather than the "boom and bust" economy. In 2012, for example, the *Sydney Morning Herald* published an article which set out the negative effects of the strong Australian dollar: a manufacturer who loses their job in the western suburbs of Sydney is unlikely to be driving a haul truck in the Pilbara (SMH 2012).

The influx of mining workers into Perth in recent times has also inflated house prices and rents. Home rental prices increased by 15% in 2012 compared to increases of 4% in other capital cities (Trenwith 2012). While this is lucrative for real estate businesses, it spells hardship for those living on a minimum wage or on a pension (ABC 2012).

By 2012 the demand for iron ore (particularly from China) had declined and the recent headlines in the online and print media were less ebullient—"Xstrata cuts 150 jobs" was one headline in 2012, for example, while in the following year "Gold producer Newmont Mining is cutting jobs at its Boddington mine in WA" (Lannin 2013) and "Newcrest Mining Cuts Jobs at Telfer, More Mines to Have Same Fate" (Su 2013).

3.4   FINAL COMMENTS

At the end of the previous chapter, we noted how incidents such as Ok Tedi, in conjunction with the global pressure on companies to be sustainable, created a shift within mining companies to be socially responsible, and to improve community engagement. The material in this chapter has demonstrated that the process of community engagement is complicated and it has also captured the breadth of views about mining within the West Australian population. It is not surprising that perceptions of mining vary; they depend on a number of factors such as place of residence, employment, ethnicity, Aboriginality, political beliefs, attitudes toward the environment and ideas about how the economy should operate.

The experiences of Aboriginal people in the Pilbara and white Australians in the country town of Boddington appear to have little in common. Colonization and racism have shaped Aboriginal identity in ways that are completely at odds with the history of white settlers, and the impact of mining on Boddington has not been as brutal as the impact on Roebourne. But what emerged from interviews for this publication, as well as a review of existing material, is that many of these views are not static, fixed, or one-dimensional: ambivalence permeates all views about mining, even for those who work at its very centre.

CHAPTER 4

Acting on Knowledge

The preceding two chapters have demonstrated how mining affects communities in different political, cultural, and social contexts. There is clearly much dissimilarity between the experiences of the Wopkaimin and the Yonggom of the Star Mountains, the Yindjibarndi people of Roebourne or the white Australian community in Boddington. Yet there are some broad general themes about the social impact of mining which emerge from this literature. None of these communities wholeheartedly endorse mining in the same way as governments and corporate elites largely because their social reality does not match the language of promise and fulfilment in corporate discourse about social responsibility and community engagement. There is also a common tension between the desire to improve individual and community livelihood from the wealth generated by mining companies and concern about the cost of participating in a mining economy. For the people of the lower Fly River, the cost has been high and has impacted on subsistence activities which are particularly important for those who cannot, or do not wish to, rely solely on a cash-based economy. For the Aboriginal people of the Pilbara, already marginalized and excluded under European rule, the sudden reversal of their position from colonial subjects to collaborators in the resource industry has been a difficult transition not least because of the expectation that they could immediately have the "capacity" to work on mine sites and mediate complicated agreements but also how to retain core cultural values while accommodating the values of the mainstream economy. Many members of the Boddington rural community, on the other hand, feel their voices are not really "heard" by the mining company because their demands—that the needs of young people should be addressed in ways other than employment, that people should be able to walk or ride their horses on local land—have no legitimacy or value.

Two questions need to be considered here: what can we learn from these narratives, and what practical value does that knowledge have for engineers and scientists who wish to situate their practice within a more just and equitable framework? There are many ways of framing the knowledge which is represented in these, and other, narratives about the social and environmental impact of mining but two key schools of thought exist:

1. Wholehearted opposition to any mining endeavor;

2. Exerting pressure on mining companies to modify their behavior.

   Some of these approaches are outlined below.

4.1 WHOLEHEARTED OPPOSITION TO MINING

4.1.1 RADICAL POLITICAL ECOLOGY AND ENVIRONMENTAL ACTIVISM: MINING AS THE BRUTAL FACE OF CAPITALISM

Political ecology is an interdisciplinary approach that acknowledges the political dimension of much recent environmental degradation and seeks to account for the "political sources, conditions and ramifications of environmental change"(Bryant 1992, p. 13). As a movement, it encompasses disciplines such as geography, anthropology, and environmental science united in a belief that there is a clear need for a "political economy of environmental conditions" particularly in the Global South, although not confined to it (Bryant 1998). Although there are a number of different approaches within political ecology, there is general agreement on a number of issues:

- So-called environmental disasters are not apolitical and these disasters historically impact on the most marginalized groups in society;

- Conflicts over access to resources must be understood within a wider context of unequal power relations not just within the state but within the global political economies in which they are embedded;

- The ideologies and goals of social movements and grassroots organizations should be taken seriously as alternatives to conventional economic models;

- Local culture, gender, and power relations influence the control of land, natural resources, labor, and capital, and mediate decision making with governments and corporations;

- Discourse theory is an important way of understanding how knowledge and power are reproduced both by governments and corporations.

Considering this range of interests, it is not surprising that many political ecologists would disagree with the claim that capitalism can be sustainably implemented (see O'Connor 1994 for example). While some political ecologists are content to depict or narrate the conditions of inequality associated with environmental degradations, others go further and argue that the world is undergoing an environmental crisis of epic proportions and that the environmental degradation which has occurred is largely a result of capitalist enterprise. They critique those who see "mining as development" as being less concerned with "saving" humanity or the planet than with "saving" capitalism. One example of such an approach would be *The Ecological Rift: Capitalism's War on the Earth* (Foster et al. 2010). The authors, who describe themselves as environmental sociologists, see all those involved in the corporate mining endeavor—as well as all social scientists—as taking a managerial approach to the environmental crisis which faces humanity. They describe this managerial approach as a form of "ecological modernisation" espoused by "green technocrats" who argue that capitalism can be sustainable (Foster et al. 2010, pp. 19–20).

While many agree with the theoretical basis of these claims they feel very little is put forward by way of practical alternatives (e.g., Walker 2006) and that there is no room at the table of radical political ecology for those who attempt to give some genuine meaning to the concept of "engagement" or who seek some justice for those who have been badly served by mining enterprises thus far.

4.1.2   GRASSROOTS MOVEMENTS AND ACTIVIST NGOS: MINING AS AN ABUSE OF HUMAN RIGHTS

There are some instances in the past—prior to the granting of Native Title—when Aboriginal groups have opposed mining ventures in Western Australia, one of the regions considered in this publication. Noonkanbah is one such example: Noonkanbah is a pastoral station in the Kimberly region to the north of the Pilbara in Western Australia. The Aboriginal Land Fund Commission purchased the pastoral lease of the Yungngora community but soon after this, the Amax Iron Ore Corporation applied for a resource tenement, that is, the right to drill an exploration well within the pastoral lease. Despite a cultural heritage report which documented "authentic and ancient" links to the site, approval was given for drilling to proceed (Ritter 2002, p. 51). Forty members of the Yungngora community locked the gates of the station and barred the way to Amax personnel and the Mines Department when they arrived in June 1979 (Ritter 2002, p. 52). In summary, the mining company was provided police protection to drill on the property in 1980 which ironically did not reveal any commercial reserves of oil. Nonetheless the act of drilling on Aboriginal land was seen as a "victory" by the Liberal state government. Richard Court, the premier, was "heavily critical of those who were seen as non-Indigenous meddlers including Senator Fred Chaney and Prime Minister Fraser, anthropologists, union protesters and the lawyers of the Aboriginal Legal Service" (Ritter 2002, p. 54). The Noonkanbah affair was a catalyst for the formation of the Kimberley Land Council and is seen as iconic in radicalizing Aboriginal politics.

Fred Chaney, who was Federal Minister for Aboriginal Affairs at the time, recollected "… at the time of Noonkanbah which was a time when the state government crushed Aboriginal opposition to drilling on the pastoral station. The company was relatively decent, Amax, and they would not have proceeded with that drilling had the state government not insisted. So even then it was the state government, rather than the miners, who were in my view behaving in a brutal fashion" (Interview 2011). This kind of activism amongst Aboriginal groups in Australia has virtually ceased since the granting of Native Title, and what is under negotiation now is the nature of the contractually binding agreements between Indigenous landowners, mining companies, and state authorities (Langton et. al 2004, O'Faircheallaigh 2012).

Grassroots movements against mining are more extensive elsewhere, and although our case studies and research is not focused on South America, it is worth noting that some observers see a correlation between poorly regulated gold mining (associated with potential toxic waste) and levels of activism in that part of the world. The largest investments in gold mining between 1990 and 2001 occurred in South America (Bridge 2004), an "increase which has led to a proliferation of environmental conflicts related to gold mining in Latin America. More than 100 conflicts of this kind are taking place in the region and most of them were initiated in the 2000s" (Urkidi 2010, p. 219).

The Latin American Mining Monitoring Program (LAMMP 2013) was formed in 1998 and claims a unique niche in the NGO sector because "it is the only non-profit, international environmental organization dedicated exclusively to the issues of women and mining, and to supporting women's Latin American organizations and Indigenous women's grass-roots groups in all stages of their activities of opposition to mining" (LAMMP 2013).

The reason for this focus is explained by one of the volunteers:

"…I realised that providing general information about companies is only important for NGOs at a certain level. Grassroots groups don't have access to internet, they don't have the capacity to understand all these issues and so we started to reflect how we can address these issues, what can we do and then also we noticed that in most events, national and international, there was very little presence of women, almost non-existent. The whole field of mining is very male-dominated. So we started looking into that and gradually we realised that women weren't even considered as stakeholders, they were ignored, that's actually a very mild word, they were excluded and NGOs made sure that women did not participate in those events. So we came up with the idea that we were going to provide support to women and see what they could do …" (LAMMP volunteer and activist, Interview 2012).

Looking at mining through the eyes of a grassroots movement such as LAMMP, a negative picture emerges in which mining companies pay lip service to the idea of community engagement by disregarding community views, marginalizing women, and silencing opposition to mining activities. Many South American women suffer a double injustice: companies who employ local men are likely to work according to pre-existing gender hierarchies in which men may have more power and status, and women are also like to suffer the impact of mining more than men (see also Lahiri-Dutt 2011a, 2011b, Macdonald and Rowland 2002).

There is consensus from some mining companies with many of these issues, particularly the gendered impact of mining (Kemp and Keenan 2009), but the implementation of real change (delivered on local terms) is unlikely to take place unless a dialog can occur between grassroots NGOs and mining corporations. This dialog is supposed to take place within a framework of stakeholder engagement but such engagement is meaningless unless the views of those most likely to be affected are actually voiced, or in this case, silenced by "big men politics" (Langton 2008).

NGOs such as LAMMP are wary of the process of "stakeholder engagement" (in South America at least) because it does not necessarily translate into stakeholder benefit:

"They [the companies] are very good at talking but the difficulty is doing the work. For example you've got Oxfam Australia, they participated in Peru in some what they call 'roundtables'. The company considers that a big success. Talk to the community; people in the community say 'that was a big mistake; we will never sit down and do roundtables with any company'. Oxfam Australia also considered that a big success, they've written books about it. People in the community say 'we thought with the support of Oxfam Australia we were going to be able to achieve something'. The company did move forward, but didn't deliver what the communities wanted" (LAMPP volunteer and activist, Interview 2012).

It appears that, unlike PNG, more local people in South America are opposed to mining because of the increased environmental degradation associated particularly with unregulated gold mining. Perhaps unsurprisingly, many South Americans see mining as another form of colonization:

"people feel that the companies are now taking up the role of governments that five hundred years ago colonized Latin America. People feel that the profit actually goes to the same powers that five hundred years ago took the resources from the countries. The difference is that five hundred years ago it was governments, colonial powers, and now it is corporations that happen to be based in those colonial powers" (LAMPP volunteer and activist, Interview 2012).

At the grassroots level—for women in South America, or for the Yonggom people on the lower Fly River, knowledge about the impact of mining is firmly based at the local, experiential level. For these people, such knowledge is not based on theories about political ecology or about Marxism but about the experience of polluted rivers, of increased violence around mine sites, or of increased commodity prices. In the vast majority of cases, the experience of local communities is not enough to halt mining activities in their areas. It is for this reason—the lack of empowerment—that many activists see mining as an abuse of human rights.

4.2 MAKING MINING COMPANIES ACCOUNTABLE

4.2.1 SELF-REGULATION: MINING COMPANIES AND SUSTAINABLE DEVELOPMENT

The Brundtland Commission famously set out a working definition of sustainable development in 1987 as "development that meets the needs of the present without compromising the ability of future generations to meet their own needs" (WCED 1987, p. 45). The Brundtland Report argued that it was "futile to attempt to deal with environmental problems without a broader perspective that encompasses the factors underlying world poverty and international inequality" (WCED 1987,

p. 3), but while the report said that sustainable development must meet all human needs (particularly those in the Global South) there are only a few paragraphs in the report which specify what those needs might be (Reid 1995, p. 57). Furthermore there was very little guidance on how the goals for action on development (WCED 1987, p. 65) should be achieved.

Social scientists and political ecologists have critiqued the concept of sustainable development (e.g., Escobar 1995, Visvanathan 1997, Sachs 1999); conservationists have argued that this definition is too limiting or too vague (IUCN 1991), and other definitions have since emerged (WWF 2002). Nonetheless the Brundtland Report signaled a significant change in thinking for both governments and corporations about the current and future impact of their activities on societies and economies.

Five years after the release of the Bruntlandt Report—and coinciding with the Ok Tedi court case against BHP in 1992—the mining industry also faced similar "pressure to improve its social, developmental and environmental performance" (Jeschke 2007). The largest of the mining transnational companies commissioned the International Institute for Environment and Development (IIED) to carry out an impartial two-year review of the mining sector measured against the global transition to "sustainable development." This project was known as "Mining and Minerals for Sustainable Development" (MMSD) and the final report—"Breaking New Ground: Mining, Minerals and Sustainable Development" (IIED 2002)—was considered a landmark review of the mining sector.

The MMSD: Mining and Minerals for Sustainable Development

While the introduction to this report asserted that mining had brought significant contributions to social and economic development it also acknowledged that it had caused significant environmental and social damage, that the benefits were less easy to discern in countries without adequate regulation, and that were many allegations about the abuse of human rights against mining companies (IIED 2002, p. 4).

The report identified a number of sustainable development principles under spheres of economy, society, environment, and governance (IIED 2002, p. xvi) and argued that implementation of these principles has to be measurable, specifically that company actions had to be monitorable, achievable and realistic (IIED 2002, p. xvii). With regard to the social impact of mining, the report identified the following concerns:

- Lack of planning about land use and land management resulting in disagreement around Indigenous land claims and protected areas;

- Mining activities do not always contribute to poverty alleviation particularly where local regulatory frameworks are not in place;

- Outsourcing of labor causes social upheaval and tension in communities; greater level of planning is required to ensure better health, education, and employment outcomes;

- More effective management of immense quantities of waste is required, and ways need to be developed of internalizing the costs of acid drainage, and impact assessments need to be improved;

- More collaboration between companies to create an integrated program of product stewardship;

- Effective public participation in decision making "requires information to be publicly available in an accessible form"; and

- Better regulatory frameworks are needed for sustainable development to be achieved.

The report specifically acknowledged that "few areas present a greater challenge than the relationship between mining companies and local communities" because of the legacy of abuse and mistrust (IIED 2002, pp. xix–xx). It assumed a "clear need" for minerals and that it was not currently possible to meet those legitimate needs without some mineral commodities in circulation. In setting out an "agenda for change" the report advocated for improvement in all the areas listed above, but what is striking about the recommendations, but perhaps understandable given that the brief for the review was set by the mining industry, is the tension between advocating a change in values and the "business case." When talking about incentives for mining companies to change, for example, the report says that "voluntary approaches are insufficient where there is a compelling priority but little or no business case to justify the expenditures needed to make it" (IIED 2002, p. xxiii).

In other words there was recognition that change would not occur unless there was a genuine commitment to do so from within a company, or "governmental intervention to achieve the same result" (IIED 2002, p. xxiii). For many activists, this raises the question of whether the desire to change is linked to a genuine desire to meet community demands or because there is a "business case" to do so. We return to this question in the section below but it should be noted for many people within the industry, good ethics also means good business.

The International Council of Mining and Minerals (ICMM)

In parallel with the MMSD project, a group of mining companies formed the International Council of Mining and Minerals (ICMM) in recognition that they were "facing significant problems in reputation, sustaining profits, access to new assets and maintaining investor and employee confidence" (ICMM 2013a). In 2001 this group of companies signed the Toronto Declaration, which signaled their intentions to implement the recommendations of the MMSD report. Implementation of those recommendations took many forms but most pertinent to our focus on social impact are the following: establishing the ten guiding principles of the ICMM Sustainable Development framework in 2003; publication of a Community Development Toolkit (ICMM 2013b) and other guidelines for community engagement; and making membership conditional on using a non-financial reporting initiative with which to measure sustainability initiatives in the social and environmental field (ICMM 2013a).

The ICMM is a voluntary initiative and membership is conditional on a) agreement to uphold the ten guiding principles of sustainable development; b) public reporting on sustainable development using the Global Reporting Initiative (GRI); and c) third party verification that the member companies are meeting the requirements of sustainable development (ICMM 2013c). Some of the 16 founding members include Anglo American, Rio Tinto, BHP Billiton, Alcoa, Noranda, Sumitomo, Mitsubishi, Lonmin, Nippon, Newmont Mining, Freeport McMoRan, and Placer Dome (Sethi and Emelianova 2006, p. 231), and there are now 22 members including, for example, AngloGoldAshanti, Barrick Gold, Codelco, Inmet, and Xstrata (ICMM 2013d).

The current president of ICMM, Anthony Hodge, explains that the procedure for joining ICMM is "stringent and diligent" and that member companies have to "commit to upholding those principles." He explained that the formation of the ICMM was:

> *"a huge step forward for the mining industry in my view and it's very creative and innovative because the companies we represent here want to make a difference. They want to be part of the solution and not just held up as being rebellious and always the cause of the problem. They truly want to be part of the solution. So that's the nature of ICMM"* (Hodge, Interview 2012).

The need for change within the culture of mining corporations is echoed by Bruce Harvey, Global Practice Leader of Communities and Social Performance at Rio Tinto who believes that qualitative data on the social and cultural context of mining is crucial:

> *"There's another whose set of data that's more qualitative, and we're talking the data that anthropologists would collect, you know—how do people live their lives around here? Where does power come from? How are decisions made? How are the checks and balances in civil society undertaken? What role do cultural or religious authorities have versus civil authorities? What are the customary norms that prevail? …. Understanding the norms of people's lives in households and in extended families is a very important part of how we will achieve societal stability in order for us to run a multigenerational mine without it becoming a victim of chaos or anarchy or civil discontent"* (Harvey, Interview 2012).

Countering this view are sceptical observers who claim that sustainable development initiatives and corporate social responsibility are a form of "greenwash." Friends of the Earth, for example, have argued that "it is often the world's most polluting corporations that have developed the most sophisticated techniques to communicate their message of corporate environmentalism" (cited in Hamann and Kapelus 2004, p. 86). There are others who agree with the need for mining companies to change their approach but feel that reporting measures need to be improved in order to be credible and effective (Azapagic 2004, Azapagic and Perdan 2010, Fonseca et al. 2012, Sethi and Emelianova 2006).

There are two areas of the ICMM sustainability codes that render them open to criticism: the codes are subjective and therefore discretionary, and the burden of proving that companies abide by their own definition of sustainable development rests primarily on an audit of internally defined criteria. This is more easily demonstrated for environmental, rather than social, performance indicators and there is increasing criticism of the "audit culture" of mining corporations which many—including business ethicists and accountants—feel is at odds with the local, complex world of social engagement. The ICMM uses a set of GRI (Global Reporting Initiative) indicators which have specially tailored for the mining industry (GRI 2010). These indicators focus on

- Labor Practices and Decent Work Performance Indicators (GRI 2010, p. 36);

- Human Rights: reporting on investment practices, non-discrimination, freedom of association, collective bargaining, child labor, and Indigenous agreements (GRI 2010, p. 38);

- Society: reporting on, for example, programs that assess the impact of mining, grievous mechanisms in place, disputes, resettlement, and mine closure plans (GRI 2010, pp. 39-41); and

- Product Responsibility (GRI 2010, pp. 41–43).

Social performance indicators may therefore include, for example, the creation of agreements, the production of social impact assessment reports (usually carried out by a third party), or reporting on the benefits to community particularly in the form of increased employment. The problem, though, is whether a tick in a box marked "social impact assessment" means that the company has responded to community expectations. Social impact assessments have been carried out in Boddington, for instance (Banarra 2012), yet the community does not feel that their voices are "heard" by the mining company. Furthermore, indicators of social performance are based on companywide performance, rather than performance at specific mine sites (Warhurst 2002, Fonseca et al. 2012, p. 8). In other words, the indicators are issue based rather than reporting on specific, local sites. There seems to be very little space in an audit culture to generate the kind of knowledge that Bruce Harvey is advocating (see interview segment above).

Despite problems with the concept of a standard audit of community engagement, these voluntary initiatives are seen as important for the member companies. Companies are aware that public opinion of the sector is low and the concept of "corporate social responsibility" is seen as serving "an important business and social purpose" because they enhance reputational effect (Sethi and Emelianova 2006, p. 228).

Reflection on these Initiatives within the Industry

A critical review of the impact of the MMSD report acknowledged that achievements have been made in understanding the complex concept of sustainable development and that ICMM has succeeded in implementing many of the original report's recommendations (Buxton 2012, p. 2). Overall, however, it was felt that while the global rules about "best practice" had been refined and improved, less attention had been paid to how those rules were implemented at the local level: "the complexity of situations at the mine site means implementation across the sector is highly variable. Questions remain as to whether current verification and reporting regimes are sufficient to meet the needs of key stakeholders—from investors to communities. In a large number of cases, there is little idea of how exactly these should be translated into progress on the ground" (Buxton 2012, p. 2).

4.2.2 NORTHERN NGOS

In this section we will consider some non-government organizations based in Europe, North America, and Australia which campaign against unjust mining practices in different ways. Mines Against Communities (MAC), for example, acts as an information hub where media reports "expose the social, economic and environmental impacts of mining, particularly as they affect Indigenous and land-based peoples" (MAC 2013). The board contains members from Australia, England, the Philippines, India, Canada, Indonesia, Peru, Chile, and Sierra Leone (MAC 2013).

MAC was formed largely in reaction to the MMSD report (see above) and to act as a critical voice to claims by mining companies that they were part of the sustainable development initiative. The report was criticized by those who felt it reflected the industry's "priority agenda of linking mining to sustainable development ... and did not reflect those of communities" and that it "failed to generate any meaningful dialogue between those most affected by mining and those most responsible" (Whitmore 2005, p. 234).

The first meeting of activists in London (chosen because it was seen as the business centre for many mining corporations) put out a formal declaration in 2001—known as the "London Declaration"—which set out a series of critiques of sustainable mining (Whitmore 2005, p. 234). The declaration stated that sustainable development initiatives from mining companies, governments, and the World Bank rested on the following half-truths or myths (MAC 2001):

- the supposed need for more and more minerals from ever more mines;

- the claim that mining catalyzes development;

- the belief that technical fix can solve almost all problems; and

- the inference that those opposed to mining mainly comprise ignorant and anti-development community and NGOs.

The London Mining Network, likewise, aims to "expose the role of companies listed on the London Stock Exchange, London-based funders and the British Government in the promotion of unacceptable mining projects" (LMN 2013).

Groups such as MAC and LMN aim to:

- Ensure that mining projects not be allowed to proceed without recognition of land title for mining-affected communities;

- Ensure that mining projects not be allowed to proceed without demonstrable public acceptance by those directly affected by them and, in the case of Indigenous Peoples, without recognition of their legal right to Free Prior Informed Consent; and

- Invite public support for campaigns to ensure that mineral development practices are consistent with goals of sustainability, human rights, and ecological justice.

Oxfam Australia, in its capacity as an advocate for human rights for local communities affected by mining, claims that, despite providing opportunities for employment and economic growth, "Australian mining companies need to show global leadership in revenue transparency and improve their practice in other countries" (Oxfam Australia 2010). Oxfam acknowledges that while "positive impacts such as employment and community development projects are important, they do not off-set the potential negatives" (Oxfam n.d.). Focusing its attention on Indonesia, Papua New Guinea, and Africa, Oxfam has identified the following real and potential negative impacts on communities in those countries:

- forcing people from their homes and land;

- preventing them from accessing clean land and water;

- impacting on their health and livelihoods;

- causing divisions in communities over who benefits from the mine and who doesn't;

- changing the social dynamics of a community; and

- exposing them to harassment by mine or government security.

Similar allegations of inconsistencies in standards have come from organizations such as Human Rights Watch (an international NGO that conducts research on, and advocacy for, human rights) who recently published a report about violence and sexual assault committed by privately contracted security forces at the Porgera mine site in PNG where Barrick Gold is the dominant shareholder (HRW 2011). The Oxfam Mining Advisory group argues that if corporations are to work in mining contexts where there is little regulation, they need to regulate their own business practices.

But how effective are local communities or NGOs in managing or resisting against mining companies that do *not* regulate their own practices, whether voluntarily or involuntarily? The actions brought against BHP by the Yonggom people of Papua New Guinea are an example of how legal action, particularly when exerted in courts of the "developed" world, can at least bring media attention to the plight of those negatively impacted by mining practices. Out-of-court settlements such as the Ok Tedi are, however, more common than judicial findings against mining companies. The European Court of Human Rights, for example, made a ruling in favor of Turkish farmers in their efforts to stop a planned gold mine, and argued in favor of the plaintiffs' argument that their right to "life and a healthy environment" would be violated if the mine were to proceed. More significantly this decision was accepted by the Turkish courts as well (MAC 2001). Such cases are, however, rare: it is more common for cases to be dismissed for lack of evidence or settled out of court. Overall the efforts of local NGOs, activist groups, and Northern NGOs have played an important role as watchdogs and in catalyzing major changes in standards of some mining companies, but it is difficult to effect real change when the interests of the state and mining companies coincide.

4.2.3   CONSENSUS BETWEEN NGOS AND MINING COMPANIES: IMPROVING MINING TECHNOLOGY

Sometimes there is agreement between NGOs (such as Mining Watch) and engineers about the need to improve current mining processes particularly in the area of waste disposal. This is evident in a recent report co-authored by Earthworks and Mining Watch Canada which details the continued destruction of environment, and thus livelihood, caused by poorly contained toxic waste disposal including, and since, the Ok Tedi disaster. The authors of this report, with input from Gavin Mudd, an environmental engineer from Australia (see Mudd 2007), claim that companies and governments must take a "precautionary approach" to tailings disposal (Cardiff et al. 2012, p. 27).

The report goes on to make the following recommendations for improvement:

- **Produce less waste.** In some cases, a mining company may prefer to build an open-pit mine that produces a large quantity of waste, but could instead build an underground mine that targets the ore more precisely and produces less waste that can be more responsibly managed. The Kemess North, now Kemess Underground, project in British Columbia, Canada, is such an example.

- **Dry stacking and backfilling.** Removing most of the water from tailings can allow mining operations to dispose of them in a dump that, if lined, covered, and reclaimed properly, is less likely to cause water contamination and threaten surrounding areas with dam failure accidents. Putting waste rock and tailings, as thickened paste, back into the pits or underground workings of mines may be a relatively responsible means of disposing of waste that reduces the mine footprint as long as operators account for pollution and accident risks.

- **Not mining.** Some places are simply not appropriate for mines. One criterion for assessing whether or not a location is appropriate for mining is if the mine operation can safely store wastes on land. Mines should not be built where it is not possible to responsibly store the waste on land. Recognizing this fact may mean less new production of some metals" (Cardiff et al. 2012, p. 28).

There are, however, many obstacles to fulfilling these recommendations. Mike Abramski, the environmental scientist who left OTML in the 1990s after the decision to use riverine disposal of mine waste, had this to say about what drove decision making in large mining corporations during his time at OTML:

> *"I think what happens in these mining companies is that you have a group of people who have this project…to get on with it, to…get this ore out, and to make a big profit for the company. And they're all…working hard in that direction. And they tend to … to egg themselves on with saying 'oh yeah we can fix that', and 'we'll fix that…later'"* (ABC 2010).

Similar views are expressed by Kapelus, who argues that mining corporations, in reality, are always trying to strike a balance between pragmatic and moral considerations: "In situations where the costs of being socially responsible do not help keep total costs down, then local managers will have to confront the tension between being socially responsible and increasing shareholder value. Here it is not as easy to facilely state that "good ethics is good business" (Kapelus 2002, p. 283).

4.3 THE GAP BETWEEN COMPANY IDEALS AND THE SOCIAL REALITY OF MINING: WHAT CAN BE DONE?

The previous two chapters depicted a change in the way mining companies interact with local communities and also in the way they apprehend "community development." While there has clearly been a change in the way mining companies operate, particularly in their negotiations with local communities (both Indigenous and non-Indigenous) there still appears be a gap between company ideals about community engagement and the reality of how communities experience the impact of mining.

A starting point for genuine dialogue on this issue is to avoid the polarizing dichotomies which can characterize debates about mining and the perceptions of those who participate in it. There are divisions within the anthropological discipline about the relative benefits of mining, and, as Ballard and Banks have noted, anthropologists have "adopted bitterly opposed stances at several mining projects" (2003, p. 306); and this is certainly true of the Ok Tedi project, as we have seen in

Chapter 2. However, Ballard and Banks argue that anthropologists should not be absolved of the "requirement for sustained reflection on the implications and consequences of our interventions" (2003, p. 306) particularly if anthropological knowledge is co-opted for the purpose of furthering corporate interests.

Few engineers are trained in critical reflection and decision making to achieve effective action for social change; consequently, engineers who are concerned about social justice may feel isolated in their profession (Riley 2008). One explanation for this, at the most general level, is that engineers are educated to be compliant and obedient to the interests of corporations and the state (Downey and Lucena 1997; Riley 2008, p. 42). Bill Townsend, who was employed by the PNG Department of Mines, and who criticized BHP for not building a tailings dam, for manipulating scientific data, and for not having any real commitment to improving the lives of local people, would be seen as an exception to the trend. But he may have been able to take this position of critical observer because he was an outsider unconstrained by any contractual obligations to the company.

In the preface we mentioned that critical understanding of core concepts in the social sciences, history, and philosophy provided a crucial basis from which to transform engineering practice, but that these concepts needed to be anchored in the pragmatic reality of ordinary life for them to have real meaning and value beyond the level of abstract debate. In the following section, we outline some of the core concepts used in corporate community engagement practices that require careful attention and critical analysis.

4.3.1   CRITICAL UNDERSTANDING OF THE LANGUAGE OF COMMUNITY ENGAGEMENT

Capacity Building

The discourse about community engagement appears, on face value, to be unassailable: its aims are worthy (to reduce poverty, to provide work) and the terminology which is used (participation and inclusivity) seems to guarantee success in partnership-building. But what are the assumptions that underlie corporate ideas about community engagement? One of the key phrases which emerges from the literature is "capacity building." Sarah Holcombe, writing about Aboriginal people in Australia, suggests we can be seduced by a language that suggests empowerment but means something quite different in reality (2006). Capacity building can therefore be synonymous with conventional ideas about development which Howitt has called "people's capacity to plan, to manage, to participate in development opportunities, to conform to the linear trajectory of rationalist development narratives" (1999, p. 5). This is explicitly recognized by Bruce Harvey, General Manager of Global Communities at Rio Tinto:

"… actually learning to drive a truck or operate some plant equipment is actually pretty easy. The thing that's really difficult for people from a pre-industrial background are the skills and attributes that we all take for granted – regularity, dependability, predictability, work in a team, sobriety, stamina, take instruction, give instruction. I mean these are the attributes that make for a regulated workforce and most people of course in many of the frontier of geographies that we're now going into have lived there according to the passage of the sun as it were and they're well adapted to that, and they live very successful lives but if you want to come and live in a mining or metals industry, you've got to have these other attributes that we learned at our grandmother's knee, we grew up learning them" (Harvey, Interview 2012).

These ideas about the "capacity" to operate in a certain economy according to certain organizational principles are culturally specific. Tonkinson (2007a) has remarked on the historical reluctance of the Mardu people of the Western Desert in the Pilbara to use "whitefella" organizing principles in the management of their own affairs but he has also seen how traditional law on who should speak and represent the Mardu community is eroded in negotiations with mining companies:

"the old true elders, the old elders who have no say now because they don't have the English, have been pushed back … they have the say in ceremonial stuff, definitely, they're still there as the elders, they still run the ceremonial things and still look after the law, in that sense but in terms of dealings with the whitefella, the outside world, particularly business dealings then they get shunted to the side because these younger people say 'we'll have to handle this. You need good English for this etc.' We've seen therefore the rise of a great deal of inequality where certain people, individuals, have amassed for themselves … have had access to a lot of money and most others have had access to very little money …." (Tonkinson, Interview 2012).

These sentiments are affirmed and reiterated by the authors of the MMSD report who later reflected on the process of writing that report:

"…not all stakeholders are organized to participate effectively on a global level … This is particularly true when talking about sustainable development, because it is so often the poor, who need to be at the centre of our concerns, who struggle to make their voices heard. It is obvious that many development problems result from trying to impose solutions without consulting the poor, who are intended as the ultimate beneficiaries… Proceeding only with those who can get to the table without help, then pretending that the resulting process speaks for everyone is not an option" (Danielson 2006, p. 11).

For engineers who are part of corporations which promote inclusive language about "cooperation, participation, ownership, multi-stakeholder dialogue, and democratic processes" the following questions should be asked: what kind of capacities are being built and for whose benefit?

Mining and Development

During the court case in which the Yonggom people sued BHP in the Victorian Supreme Court, Kirsch (1997) documented one of the lawyer's remarks to the judge on what "loss of amenity" meant to the Yonggom people:

> *"these plaintiffs are people who live a subsistence lifestyle. They live substantially, if not entirely, outside the economic system which uses money as the medium of exchange. But to say that does not alter the fact that if they are deprived of the very things which support their existence, they suffer loss. Of course it is a loss which appears in an uncommon guise because typically the courts have dealt with claims that are rooted in society's adherence to the monetary medium of exchange* (Victorian Supreme Court, 14 October 1995:58, cited in Kirsch 1997).

> *It simply cannot be right that because people exist outside the ordinary economic system, they therefore do not have rights where their lives are damaged by the negligence of others. Now the lifestyle of the Papua New Guinea natives in gathering food, fishing and game and the like and using it to eat or sell is no less an economic activity because it is not translated through the medium of money. It is economic loss to be deprived of your source of food ... whether measured in money or not (ibid:59-60)"* (cited in Kirsch 1997, p. 135–6).

Kirsch cited the passage to underline how novel this idea appeared to the courts of law in Australia, and how this could provide an important legal precedent for the adjudication of other claims in the future. For our purposes, however, this excerpt has even greater significance: the recognition that other economies have equal value is of fundamental importance to our critical understanding of "development."

What we have found in this area is a similar pattern of disjuncture between global discourse and local experience. Mining companies, such as Rio Tinto (see Chapter 3) tend to view development in conventional terms, that is, success can be measured by the provision of work and by employment indicators. This view is also shared by many Aboriginal people. In an interview with Barry Taylor, an Aboriginal entrepreneur in the mining sector, he said that work was the only way out of welfare dependency, and the only way forward for his people. Yet other people such as mine managers working directly with Aboriginal people, or Aboriginal people living in remote areas, realize that it is possible to accommodate different types of economies. Not all local people—Indigenous or white—will want to work on a mine site and do not necessarily measure progress in monetary terms.

The main point here is to recognize that there is "inevitable contestation over economic values" and that not everyone will "meekly acquiesce to some predetermined pathway to modernity proposed for them" (Altman 2012, p. xiv). While the development discourse promises a great deal, it is important to remember that "the mining town frequently functions as a symbol and promise of modernity for local communities and workers alike, though residents all too frequently find themselves betrayed, cast aside and disconnected from the processes of development and modernity that globalization promises" (Ballard and Banks 2003, p. 292).

4.3.2   PRACTICAL ALTERNATIVES TO THE AUDIT CULTURE OF COMMUNITY ENGAGEMENT

We mentioned earlier that a number of observers, ranging from accountants to anthropologists, have expressed concern about the "audit culture" within mining corporations, particularly with regard to measuring their success in community engagement. Kemp et al. (2012) have summarized these concerns as follows: using a checklist method to demonstrate corporate social responsibility is an instrumentalist approach which means that people become subjects rather than participants in the process of community engagement; and the audit process constrains the opportunity for critical reflection and dialog.

Self-reflexivity and critical understanding of key issues is, we argue, a necessary basis for practical action. As Kemp et al. have argued, "a high-level structural analysis does not necessarily provide insight into problems and possibilities of community development in mining that involves impacted and affected families, households, small groups and local communities. A practice-based focus has the potential to improve understanding of mining and development at the local level and enrich broader debates" (2012, pp. 1– 2).

Engaging with Community

Kemp et al. (2012) recommend a series of alternatives to the current use of company-wide indicators that tend to obscure the differences in context between mine sites which are owned by the same company (such as that even within Western Australia or between Western Australia and Papua New Guinea):

- **Facilitated self-assessment:** whereby a member of staff from the operational level is invited to co-author an assessment report and to consult with local community representatives to get a richer sense of community expectations, to be critically self-reflective about their own views; and to create a space for dialog (Kemp et al. 2012, p. 6).

- **Practice clinics:** whereby attention to the personal experience of mining practitioners is the focus point—rather than documenting performance or measuring dissonance—and where an experienced facilitator enables participants to imagine solutions that are aligned with their own "ethical principles and moral norms." The starting point of these clinics is the local rather than the global, and based on personal experience rather than corporate discourse. The overall aim is to use "creative reasoning around sensitive, complex and particular practice dilemmas" (Kemp et al. 2012, p. 6).

- **Organizational ethnography:** whereby organizational life is observed over an extended period to understand what shapes corporate response to local incidents and expectations. This approach requires a long-term interdisciplinary approach which is more costly; for these reasons, the authors state that "this can be challenging for companies who favor rapid, broad-brush approaches to social audit and assessment for their cost-effectiveness. The other two methods canvassed above can provide more immediate and demonstrable value to the company, but it may take weeks or even months before ethnographic methods produce data that would be considered 'useful' from a company perspective" (Kemp et al. 2012, p. 7). Nonetheless, many mining companies do employ anthropologists as consultants but little research has been done on whether the reports or articles which they produce do in fact impact on company policy or practitioner's behavior.

Currently, there is little need for engineers to "engage with communities" because this work is done for them by specialists in the field such as sociologists, anthropologists, or people with professional accreditation in the field of social impact. Yet in our interviews a number of people spoke about the need for engineers to have "direct contact" with communities particularly in the Indigenous context. Janina Gawler, General Manager of Communities at Rio Tinto (Pilbara Iron), commented that most people in Australia "go through life without any direct contact with Aboriginal people" yet when engineers start work on a mine site, they are working on other people's country and "peering at a power point" does not convey the nature of Aboriginal ties to the land, nor about the impact of their history (Gawler, Interview 2011).

In our interviews with Fred Chaney he also made the comment that "there is no substitute for contact" and spoke approvingly of the way in which many transnational companies such as Rio Tinto, as well as national banks and non-mining ventures, took their senior management to rural and remote locations to camp and speak with Aboriginal people. There are, however, a number of problems with this approach. The inspiration which senior managers may feel from hearing about community needs and expectations at first hand does not necessarily trickle down to lower level employees. Multinational mining companies are also very diverse in their approach to "community engagement" but to date there has been little or no ethnographic research on the organizational culture of mining companies and what individual engineers may feel about the values of the corporations for whom they work.

4.4   A RETURN TO THE BEGINNING: WHAT ENGINEERS NEED TO KNOW

In the preface to this publication, we indicated that one of our key questions to those working within and without the mining industry was "what do engineers need to know?" in order to not repeat the mistakes of the past, and to create an engineering practice that is informed, not just by principles of efficiency or technical innovation, but by a critical awareness of social impact. Here are three responses to those questions:

"It is part of your job as engineers to ensure that you're good not only at roads and dams and bridges and structures, but that you understand that those things impact on the lives of the people who live where you're working, and perhaps even people a long way away. So this is an intricate part, an integral part of what you are doing when you carry out engineering works which you might once have been taught were just matters of mathematics and stresses and tensions and so on, and tolerances. This is about things which impact on human populations" (Chaney, former Federal Minister for Aboriginal Affairs, Board member of Reconciliation Australia, Interview 2011).

"The skills sets (required to deal with people in the developing world) will be quite different and the capacity around engagement with communities, human rights, ethics, are all fundamental whether you are in Australia or whether you're in a developing country and if you cannot operate in those environments you put your business and yourself at great risk, and having some capacity to deal with those challenges requires some flexibility and comprehension of the impact. Unless there's some interface that acknowledges these issues, then you will end up with problems in the long term. So from an engineering point of view or a project or study team point of view, they have to be taken into consideration and be given fair voice and a decent result because if you don't take care of that then your perfect engineering piece won't succeed, and a notion about delivering on time, under budget, has to be in the context of the environment in which you're working and that's both the physical and social environment and if that's not addressed, then you don't have the capacity as far as I'm concerned" (Janina Gawler, General Manager of Communities, Rio Tinto Pilbara Iron, Interview 2011).

"… what our mining companies need most now are people that know how to build relationships. But you have to do that with the knowledge of the mining process. You have to have a fundamental understanding of engineering but you have to be able to respect people, you have to understand how to interact effectively with people because companies are working in different cultures. Our CEO said that the greatest insurance policy is trust with our communities and you cannot build that trust by being a paternalist bugger … in order to build relationships you've got to know how to do it. You've got to have the skill at it, the sensitivity of it. Have to learn just not to listen, but hear and those are skills that are not typical of our engineering faculties. So how does it change?" (Anthony Hodge, President of the International Council of Mining & Minerals, Interview 2012).

Putting these ideals into practice is certainly difficult. Mining corporations are vast enterprises employing many people and tendering out their work to many more. There are many different corporate cultures *within* a mining venture particularly when there are "complex webs of subsidiaries and shared project ownership" (Ballard and Banks 2003, p. 293). The fluid and complex nature of corporate capital precludes, to a large extent, the fostering of a shared ethos or shared concerns about the social impact of professional practice. The division of labor that is essential to this system also means that, to date, engineers are employed to solve and implement the technical components while other professionals are employed to deal with the social impact of mining.

Our goal is to broaden this narrow focus of engineering work and to provide some means by which this can be achieved. This is both a modest and significant claim. The anarchist geographer Elise Recluse wrote that: "great enthusiasm and dedication to the point of risking one's life are not the only way of serving a cause… the conscious revolutionary is not only a person of feeling, but also one of reason, for whom every effort to promote justice and solidarity rests on precise knowledge …" (cited in Harvey 2008, p. ix). The ability to question the paradigms in which we work rests on the kind of knowledge to which Recluse refers.

Narratives about the impact of mining comprise a complex and often contradictory body about knowledge which make it difficult to draw precise conclusions. It would not be accurate, for example, to claim that all Indigenous people oppose mining because it clashes with their cultural ideas about nature; it is also imprecise to claim that mining is axiomatic to civilization and progress when there is staggering evidence to the contrary. Knowledge that reflects social reality is, therefore, troublesome in that it does not provide neat solutions to pressing questions on how to best alleviate injustice and inequality.

We, the authors of this text, work in a context in Western Australia in which mining has been integrated with development since the first days of settlement. Its history is a complex blend of boom, bust, hardship, and economic development, the promise of a future for many and the demise of a culture for others. How can anyone get their head around this technological/social/economic and environmental melting pot? What we have attempted to do in this short publication, along with the others in this series, is to raise awareness of a set of critical questions that might be asked of any present endeavor. We insist on questioning assumptions, particularly those which remain unquestioned, as they are part of what we understand as "common sense" because they are all around us in everyday discourse—conversation, patterns of thinking, and acting. We thus question things which seem evident and normal. Who does benefit from mining practices? Why? How might these benefits be better distributed? How can the costs be limited or even eliminated? We have adopted a social and environmental justice lens; we have questioned and critiqued and tried to see many different local and global viewpoints that perhaps do not usually get considered.

The knowledge which is presented in this publication will, we hope, provide a platform from which engineers can better understand the economic imperatives which shape corporate behavior, and to be better able to understand the potential social impact of their work at the local level. We aim to provide a wide range of viewpoints on the actions and activities of engineers and the skills and ways of thinking to accompany these, which will enable future engineers to be part of a broader, interdisciplinary and globally interrelated set of solutions to some of our more intractable social and environmental issues.

References

Act Now! (2013). Act Now! For a Better Papua New Guinea. Retrieved from: http://www.actnow-png.org/

ABC Australian Broadcasting Corporation. (2010). *A Dirty Business*. Reporter: Andrew Fowler, 4 Corners documentary. Retrieved from: http://www.abc.net.au/4corners/content/2010/s2867659.htm

ABC Australian Broadcasting Commission. (2012). Low Income Earners Squeezed out of Rentals (April 30 2012). Retrieved from: http://www.abc.net.au/news/2012-04-30/low-income-earners-squeezed-out-of-rentals/3979962

ABS Australian Bureau of Statistics. (2010). National Regional Profile: Boddington (LGA). Retrieved from: http://www.abs.gov.au/AUSSTATS/abs@.nsf/Previousproducts/LGA50630Industry12004-2008?opendocument&tabname=Summary&prodno=LGA50630&issue=2004-2008

ABS Australian Bureau of Statistics. (2011). National Regional Profile: Boddington. Retrieved from: http://www.abs.gov.au/AUSSTATS/abs@nrp.nsf/Previousproducts/LGA50630Economy12006-2010?opendocument&tabname=Summary&prodno=LGA50630&issue=2006-2010

ABS Australian Bureau of Statistics. (2013). WA Experiencing Record Population Growth as Australia Approaches 23 Million. Retrieved from: http://www.abs.gov.au/ausstats/abs@.nsf/latestProducts/3101.0Media%20Release1Sep%202012

ABS Australian Bureau of Statistics. (2013a). National Regional Profile: Boddington Population. Retrieved from: http://www.abs.gov.au/AUSSTATS/abs@nrp.nsf/Previousproducts/LGA50630Population/People12006-2010?opendocument&tabname=Summary&prodno=LGA50630&issue=2006-2010&num=&view=

Adkhikari, A., Sen, A., Brumbaugh, R., and Shwartz J. (2011). Altered Growth Patterns of a Mountain Ok Population of Papua New Guinea Over 25 Years of Change. *Amer J Human Biol* 23:325–332. DOI: 10.1002/ajhb.21134.

Agricola, G. (1912). *De Re Metallica*. Translated from the first Latin edition of 1556 by Herbert Clark Hoover & Lou Henry Hoover. London: Mining Magazine.

AICD Australian Institute of Company Directors. (2006). The Good Corporate Citizen. Retrieved from: http://www.companydirectors.com.au/Director-Resource-Centre/Publications/Company-Director-magazine/2000-to-2009-editions/2006/March/cover-story-the-good-corporate-citizen

AIHO Australian Indigenous Health Infonet. (2012). Population Structure: Western Australia. Retrieved from: http://www.healthinfonet.ecu.edu.au/states-territories-home/wa/reviews/our-review/population-structure

AIHW Australian Institute of Health and Welfare (2013). Indigenous Health: Overview. Retrieved from: http://www.aihw.gov.au/indigenous-health/

Altman, J. (2005). Development Options on Aboriginal land: Sustainable Indigenous Hybrid Economies in the Twenty-First Century. In L. Taylor, G. K. Ward, G. Henderson, R. Davis and L.A. Wallis (eds), *The Power of Knowledge: The Resonance of Tradition*. Canberra: Aboriginal Studies Press.

Altman, J. (2009). Contestations over Development. In J. Altman and D. Martin (Eds), *Power, Culture, Economy: Indigenous Australians and Mining* (pp. 1-16). CAEPR (Centre for Aboriginal Economic Policy Research) Research Monograph No. 30. Canberra: ANU e-Press.

Altman, J. (2012). Foreword. In N. Fijn, I. Keen, C. Lloyd, and M. Pickering (Eds), *Indigenous Participation in Australian Economies II: Historical Engagements and Current Enterprises*. Canberra: ANU ePress.

Altman, J. and Martin, D. (Eds). (2009). *Power, Culture, Economy: Indigenous Australians and Mining*. Canberra: Centre for Aboriginal Economic Policy Research (CAEPR), ANU.

Armstrong, R. and Baillie, C. (2012). Engineers Engaging with Community: Negotiating Cultural Difference on Mine Sites. *Intl J Eng, Soc Justice and Peace*, 1(2): 7-17.

Australian Human Rights Commission. (2009). The Australian Mining and Resource Sector and Human Rights. Retrieved from: http://humanrights.gov.au/pdf/education/business_and_human_rights/factsheet3.pdf

Auty, R. (1993). *Sustaining Development in Mineral Economies: The Resource Curse Thesis*. London: Routledge. DOI: 10.4324/9780203422595.

Azapagic, A. (2004). Developing a Framework for Sustainable Development Indicators for the Mining and Minerals Industry. *J Cleaner Prod*, 12: 639–662. DOI: 10.1016/S0959-6526(03)00075-1.

Azapagic, A. and Perdan, S. (2010). *Sustainable Development in Practice: Case Studies for Engineers and Scientists*. Hoboken: Wiley. DOI: 10.1002/9780470972847.

Baillie, C. and Catalano, G. (2009). *Engineering and Society: Working Toward Social Justice, Part I: Engineering and Society*. Synthesis Lectures on Engineering, Technology & Society. San Rafael, CA: Morgan & Claypool Press. DOI: 10.2200/S00136ED1V01Y200905ETS008

Bakewell, P. (1984). *Miners of the Red Mountain : Indian Labor in Potosi, 1545-1650.* Albuquerque: University of New Mexico Press.

Ballard, C. and Banks, G. (2003). Resource Wars: The Anthropology of Mining. *Annual Rev of Anthropol* 32: 287-313. DOI: 10.1146/annurev.anthro.32.061002.093116.

Bamblett, M. Harrison, J., and Lewis, P. (2010). Proving Culture and Voice Works: Toward Creating the Evidence Base for Resilient Aboriginal and Torres Strait Islander Children in Australia. *Int J Child and Family Welfare*, 1-2, 98-113.

Banarra. (2012). "Newmont Boddington Gold: Social Impact Assessment Summary Report". Report published for Newmont Boddington Gold.

Banks, G. (1993). Mining Multinationals and Developing Countries: Theory and Practice in New Guinea. *Applied Geo*, 13: 313-327. DOI: 10.1016/0143-6228(93)90035-Y.

Banks, G. (2002). Mining and the Environment in Melanesia: Contemporary Debates Reviewed. *The Contemporary Pacific*, 14(1): 39-67. DOI: 10.1353/cp.2002.0002.

Banks, G. (2009). Activities of TNCs in Extractive Industries in Asia and the Pacific: Implications for Development. *Transnational Corp*, 18(1): 43-60.

Barber, M. and Jackson. S. (2011). *Water and Indigenous People in the Pilbara, Western Australia: A Preliminary Study.* CSIRO Water for a Healthy Country Flagship. Perth: CSIRO.

Barber, M. and Jackson. S. (2012). Aboriginal Water Values and Resource Development Pressures in the Pilbara Region of North-West Australia. *Australian Aboriginal Studies*, 2: 32-49.

Bell, F. G and Donnelly, J. J. (2006). *Mining and Its Impact on the Environment.* London: Taylor & Francis.

Benson, P. and Kirsch, S. (2010). Capitalism and the Politics of Resignation. *Current Anthropol*, 51(4), p.459-486. DOI: 10.1086/653091.

BHPBilliton. (2002). BHP Billiton Withdraws from Ok Tedi Mine and Establishes Development Fund for Benefit of Papua New Guinea People. Retrieved from: http://www.bhpbilliton.com/home/investors/news/Pages/Articles/BHP%20Billiton%20Withdraws%20from%20Ok%20Tedi%20Copper%20Mine%20and%20Establishes%20Development%20Fund%20for%20Benefit%20of%20Papua%20New%20Guinea.aspx

Bindler, R., Renberg, I., Rydberg, J., and Andrén, T. (2009). Widespread Waterborne Pollution in Central Swedish Lakes and the Baltic Sea from Pre-Industrial Mining and Metallurgy. *Environmental Pollution*, 157(7): pp.2132-41. DOI: 10.1016/j.envpol.2009.02.003.

Blanchard, I. (1998). The Long Sixteenth Century. Conference paper presented to the Twelfth International Congress, Madrid (August). Retrieved from: http://www.ianblanchard.com/CEU/paper3.pdf

Bolton, G.C. (1981). Black and White after 1897. In C. T Stannage (Ed.), *A New History of Western Australia*. Perth: University of Western Australia Press.

Bolton, G. C. (2008). *Land of Vision and Mirage: Western Australia since 1826*. Perth: University of Western Australia Press.

Bosworth, M. and Jean, A. (1999). Interpreting confinement in North-Western Australia. *Historical Env*, 14(2): 25-32.

Botin, J. A. (Ed.). (2009). Sustainable Management of Mining Operations. Colorado: Society for Mining, Metallurgy and Exploration (SME).

Boyle, S. (2002). Resource Development and Health in Papua New Guinean Ok Tedi Mines. *Rural and Remote Environmental Health*, 1: 62-65. Also available from: http://www.tropmed.org/rreh/vol1_8.htm

Braudel, F. (1982). *The Wheels of Commerce: Civilization & Capitalism 15th–18th Century*, Volume 2. New York: Harpers & Row.

Bridge, G. (2004). Mapping the Bonanza: geographies of mining investment in an era of neoliberal reform. *The Professional Geographer*, 56 (3): 406–421.

Brock, P. (2004). Skirmishes in Aboriginal History. *Aboriginal History*, 28: 207-225.

Brueckner, N. , Durey, A., Mayes, R., and C. Pforr. (2013). The Mining Boom and Western Australia's Changing Landscape. Toward Sustainability or Business as Usual. *Rural Soc J*. 22(2), 111-124. DOI: 10.5172/rsj.2013.22.2.111.

Bryant, R.L. (1992). Political ecology: an emerging research agenda in third world studies. *Political Geography*, 11: 12-36.

Bryant, R. (1998). Power, Knowledge and Political Ecology in the Third World: A Review. *Progress in Physical Geography*, 22(1): 79-94. DOI: 10.1177/030913339802200104.

Burbank, V.K. (2006). From Bedtime to On Time: Why Many Aboriginal People Don't Especially Like Participating in Western Institutions. *Anthropological Forum*, 16(1), 3-20. DOI: 10.1080/00664670600572330.

Burbank, V. K. (2011). *An Ethnography of Stress: The Social Determinants of Health in Aboriginal Australia*. New York: Palgrave Macmillan. DOI: 10.1057/9780230117228.

Burawoy, M. (1998). The Extended Case Method. *Sociological Theory* 16(1): 4-33. DOI: 10.1111/0735-2751.00040.

Burton, J. (1997). Terra Nugax and the Discovery Paradigm: How Ok Tedi Was Shaped by the Way It Was Found and How the Rise of Political Process in the North Fly Took the Company by Surprise. In G. Banks & C. Ballard (Eds.), "The Ok Tedi Settlement: Issues, Outcomes and Implications," pp. 27-55. Canberra: National centre for Development Studies, Pacific Paper No. 27, RSPAS, Australian National University.

Burton, J. (2000). Knowing about Culture: The Handling of Social Issues at Resource Projects in Papua New Guinea. In A. Hooper (Ed.), *Culture and Sustainable Development in the Pacific*, pp. 98-110. Canberra: Asia Pacific Press (also available now on ANU E-press).

Buxton, A. (2012). *MMSD + 10: Reflecting on a Decade of Mining and Sustainable Development.* London: IIED.

Capitalco (n.d.). *Iron Ore Mining in Western Australia.* Retrieved from: http://www.capitalco.com.au/Portals/0/Docs/Mining_Resources/Iron%20Ore%20Mining%20in%20Western%20Australia.pdf

Cardiff, S., Coumans, C., Hart, R., Sampat S., and Walker, W. (2012). *Troubled Waters: How Mining Waste is Poisoning Our Oceans, Rivers and Lakes.* Ottawa: EarthWorks and Mining Watch Canada.

Carter, F. W. (2006). *Trade and Urban Development: An Economic Geography of Cracow from its Origins till 1795.* Cambridge: Cambridge University Press.

Chapman, P., Burchett, M, Campbell, P., Dietrich, W. and Hart, B. (1997). Fourth Report of the OTML Environment Peer Review Group. Retrieved from: http://www.oktedi.com/attachments/241_MWMP_Comment_on_Sci.pdf

Chapman, P., Burchett, M, Campbell, P., Dietrich, W. and Hart, B. (2000). Comments on Key Issues and Review Comments on the Final Human and Ecological Risk Assessment Documents. Retrieved from: http://www.oktedi.com/attachments/244_MWMP_PRG_FinalReport.pdf

Copper Country History. (n.d.). *One-Man Drill.* Retrieved from: http://coppercountry.wordpress.com/tag/one-man-drill/

Cordero, H. and Tarring, L. (Eds). (1960). *From Babylon to Birmingham: An Historical Survey of the Development of the World's Non-Ferrous Metal and Iron and Steel Industries and of the Commerce in Metals Since the Earliest Times.* London: Quin Press.

Coronado, G. and Fallon, W. (2010). Giving with One Hand: On the Mining Sector's Treatment of Indigenous Stakeholders in the Name of CSR. *Int J Sociol and Soc Policy*, 30 (11/12): 666-682. DOI: 10.1108/01443331011085259.

Craddock, P. (1995). *Early Metal Mining and Production.* Edinburgh: Edinburgh University Press.

Crook, A. (2013). Marcia Langton Sparks Academic Spat over Charges of "Racism." 27 February, Crikey. Retrieved from: http://www.crikey.com.au/2013/02/27/marcia-langton-sparks-academic-spat-over-charges-of-racism/?wpmp_switcher=mobile

Danielson, L. (Ed.) (2002). Breaking New Ground: Mining, Minerals and Sustainable Development. London: IIED (International Institute for Environment and Development). Retrieved from: http://www.iied.org/mmsd-final-report

Danielson, L. (2006). Architecture for Change: An Account of the Mining, Minerals and Sustainable Development Project History. Berlin: Global Public Policy Institute. Retrieved from: http://pubs.iied.org/pdfs/G00975.pdf?

Decker, J. L. (1997). *Made in America Self-Styled Success from Horatio Alger to Oprah Winfrey*. Minneapolis: University of Minnesota Press.

De Bruyn, I. A. and Bell, F. G. (2001). The Occurrence of Sinkholes and Subsidence Depressions in the Far West Rand and Gauteng Province, South Africa, and their Engineering Implications. *Env Eng Geosci*, 7(3): 281-295.

Department of Conservation and Environment. (1985). "Worsley Alumina Joint Venturers Boddington Gold Mine Proposal: Environmental Protection Authority Report and Recommendations." Perth, Western Australia: Department of Conservation & Environment.

Department of Local Government and Regional Development. (2003). *Indicators of Regional Development in Western Australia*. Retrieved from: http://myweb.westnet.com.au/bobcocking/pdfs/Indicators.pdf

DMP (Department of Mines & Petroleum). (2011). WA Minerals and Petroleum Statistics Digest. Perth: Government of Western Australia. Also available at: http://www.dmp.wa.gov.au/documents/121857_Stats_Digest_2011.pdf

Department of Regional Development & Lands. (2011). *Regional Centres Development Plan: Background Information*. Retrieved from: http://www.drd.wa.gov.au/publications/Documents/Regional_Centres_Development_Plan_SuperTowns.pdf

Department of Water. (2010). *Pilbara Regional Water Plan 2010-2030*. Perth: Department of Water.

Dore, E. (2000). Environment and Society: Long-Term Trends in Latin American Mining. *Env and Hist*, 6(1): 1-29. DOI: 10.3197/096734000129342208.

Doohan, K. (2008). *Making Things Come Good: Relations between Aborigines and Miners at Argyle*. Broome, WA: Backroom Press.

Downey, G.L and J.C. Lucena. (1997). Engineering Selves: Hiring in to a Contested Field of Education. In G.L. Downey and J Dumit (Eds) *Cyborgs and Citadels*, pp. 117-141. Sante Fe, New Mexico: School of American Research Press.

Duncan, L. C. (1999). Roman Deep Vein Mining. Retrieved from: http://www.unc.edu/~duncan/personal/roman_mining/deep-vein_mining.htm

Edmondson, J. C. (1989). Mining in the Later Roman Empire and Beyond: Continuity or Disruption? *J Roman Studies*, 79: 84-102. DOI: 10.2307/301182.

Edmunds, M. (1989). *They Get Heaps: A Study of Attitudes in Roebourne, Western Australia*. Canberra: Aboriginal Studies Press.

Edmunds, M. (2012a). Doing Washing in a Cyclone, or a Storm in a Teacup? Aboriginal People and Organizations in the Pilbara: An Overview. In B. Walker (Ed.), *The Challenge, Conversation, Commissioned Papers and Regional Studies of Remote Australia*. Alice Springs: Desert Knowledge Australia.

Edmunds, M. (2012b). Roebourne: A Case Study. In B. Walker (Ed.), *The Challenge, Conversation, Commissioned Papers and Regional Studies of Remote Australia*, pp. 152-180. Alice Springs: Desert Knowledge Australia.

ECS (Economic Consulting Services). (2004). "Water and the Western Australian Minerals and Energy Industry: Certainty of Supply for Future Growth." Report prepared for the Chamber of Minerals and Energy of Western Australia. Retrieved from: http://www.ret.gov.au/resources/documents/industry%20consultation/regional%20minerals%20program/water%20and%20the%20west%20australian%20minerals%20and%20energy%20industry/rmp_water_and_the_west_australian.pdf

ECS (Economic Consulting Services). (2008). "Needs Analysis for the Shires Impacted by the Re-Opening of the Boddington Gold Mine." Report prepared for the Department of Industry and Resources. Perth: ECS.

EPA (Environmental Protection Agency). (1994). "Boddington Gold Mine: Rehabilitation Strategy: Worsley Alumina Pty Ltd: Proposed Changes to Environmental Conditions: Report and Recommendations of the Environment Protection Authority." Perth, Western Australia: Environmental Protection Agency.

EPA (Environmental Protection Agency). (2001). "Boddington and Hedges Gold Mines, Boddington Expansion, Shire of Boddington, Change to Environmental Conditions, and Gas-Fired Power Station and Natural Gas Pipeline 12 km Northwest of Boddington." Perth, Western Australia: Environmental Protection Agency.

EPA (Environmental Protection Agency). (2002). "Telfer Project, Expansion of Telfer Gold Mine, Great Sandy Desert: Report and Recommendations of the EPA." Perth, Western Australia: EPA.

Escobar, A. (1995). *Encountering Development: The Making and Unmaking of the Third World*, Princeton, N.J.: Princeton University Press.

Esteves, A. H., Brereton, D., Samson, D., and Barclay, M. A. (2010). *Procuring from SMEs in Local Communities. A Good Practice Guide for the Australian Mining, Oil and Gas Sectors*. Brisbane: Centre for Social Responsibility in Mining, UQ.

Evans-Pritchard, E. (1950). Social Anthropology: Past and Present. *Man*, 50:118-124.

Filer, C, (1990). The Bouganville Rebellion, the Mining Industry and the Process of Social Disintegration in Papua New Guinea. *Canberra Anthropology*, 13 (1), 1-39. DOI: 10.1080/03149099009508487.

Filer, C. and Imbun, B. (2009). A Short History of Mineral Development Policies in Papua New Guinea, 1972-2002. In R. J.May (Ed.), *Policy Making and Implementation: Studies from Papua New Guinea*, pp. 75-116. Canberra: ANU E-press.

Filer, C. and Macintyre, M. (2006). Grass Roots and Deep Holes: Community Responses to Mining in Melanesia. *Contemporary Pacific*, 18(2), 215-227. DOI: 10.1353/cp.2006.0012.

Ferguson, J. (2005). Seeing Like an Oil Company: Space, Security and Global Capital. *Amer Anthropol* 107(3): 377-382. DOI: 10.1525/aa.2005.107.3.377.

Fogarty, P. (2012). Residents Priced out of Australia's Powerhouse Port. September 5, British Broadcasting Commission. Retrieved from: http://www.bbc.co.uk/news/world-asia-19222035

Fogarty, P. (2012). Good Times for All in Boomtown Perth? August 28, 2012. British Broadcasting Commission. Retrieved from: http://www.bbc.co.uk/news/world-asia-19222037

Fonseca, A., McAllister, M., and Fitzpatrick, P. (2012). Sustainability Reporting Among Mining Corporations: A Constructive Critique of the GRI Approach. *J Cleaner Prod*, pp. 1-14. DOI: 10.1016/j.jclepro.2012.11.050.

Foster, J. , Clark, B., and York, R. (2010). *The Ecological Rift: Capitalism's War on the Earth*. New York: Monthly Review Press.

Fox, L. (2013). Trust Loses Challenge against PNG's Ok Tedi Mine Takeover. October 2, Australian Broadcasting Commission. Retrieved from: http://www.abc.net.au/news/2013-10-02/an-png-court-upholds-govt-ok-tedi-takeover/4994086

Frankel, B. (2013). "Mining Does Not Offer the Answer to Indigenous Woes." March 12, Sydney Morning Herald. Retrieved from: http://www.theage.com.au/comment/mining-does-not-offer-the-answer-to-indigenous-woes-20130311-2fw7s.html

Franklin, U. (1990). *The Real World of Technology*. Concord, Ontario: Anansi Press.

Furniss, E. (2005). Imagining the Frontier: Comparative Perspectives from Canada and Australia. In D. Bird-Rose (Ed.) *Dislocating the Frontier: The Mystique of the Outback*, Chapter Two. Canberra: ANU e-Press. Retrieved from: http://epress.anu.edu.au/dtf/mobile_devices/ch02.html

Jackson, R. (2003). "Muddying the Waters of the Fly: Underlying Issues or Stereotypes?" Resource Management in Asia-Pacific Working Paper No. 41, Research School of Pacific & Asian Studies. Canberraf: ANU.

Garrett, G. and Garrett, N. (2013). Copper Mining Strike of 1913. Department of Natural Resources. Retrieved from: http://www.michigan.gov/dnr/0,4570,7-153-54463_19313_20652_19271_19357-156714--,00.html

Gawler, J. and Harvey, B. (2005). Emerging Models of Diversity in Australian Mining. Center for Social Responsibility in Mining. The Fifth International Conference on Diversity in Organizations, Communities and Nations, Beijing, China, (30th June to Sunday 3rd July).

Geoscience Australia (2012). Copper: Fact Sheet. In *Australian Atlas of Minerals Resources, Mines and Processing Centers, Department of Resources, Energy & Tourism*. Retrieved from: http://www.australianminesatlas.gov.au/education/fact_sheets/copper.html

Gibson, G., MacDonald, A., and O'Faircheallaigh, C. (2011). Cultural Considerations for Mining and Indigenous Communities. *SME Mining Engineering Handbook*. Third Edition Ed. P. Darling. (Volume 2). (pp 1797- 1816)Society for Mining, Metallurgy, and Exploration, Inc.

Gilberthorpe, E. (2013). Community Development in Ok Tedi, Papua New Guinea: The Role of Anthropology in the Extractive Industries. *Comm Devel J*, 48(3): 466–483. DOI: 10.1093/cdj/bst028.

Gilberthorpe, E. and Banks, G. (2012). Development on Whose Terms? CSR Discourse and Social Realities in Papua New Guinea's Extractive Industries Sector. *Resources Policy*, 37: 185-193. DOI: 10.1016/j.resourpol.2011.09.005.

Godden, L., Langton, M., Mazel, O., and Tehan, M. (2008). Introduction. In Accommodating Interests in Resource Extraction: Indigenous Peoples, Local Communities and the Role of Law in Economic and Social Sustainability, *J Energy & Nat Res Law* Special Issue, 26(1): pp. 1-30.

Gottlieb. R. (2010). Ecocide. In W. Jenkins and W. Bauman (Eds). *The Spirit of Sustainability*. Barrington, MA: Berkshire Publishing Company.

Grabosky, P. (1989). Wayward Governance, Illegality and Its Control in the Public Sector. Canberra: Australian Institute of Criminology. Retrieved from: http://www.aic.gov.au/publications/previous%20series/lcj/1-20/wayward/ch5t.aspx

Graeber, D. (2001). *Toward an Anthropological Theory of Value: the False Coin of Our Own Dreams.* New York: Palgrave. DOI: 10.1057/9780312299064.

Graham, D. (n.d.). *WA Blacks Make Huge Land Claim.* The Age. Retrieved from: http://www.mabonativetitle.com/info/WABlacksMakeHugeLandClaim.htm

Graulau, J. (2008). Is Mining Good for Development? *Prog Devel Studies*, 8(2): 129-162. DOI: 10.1177/146499340700800201.

Green, N. (1981). Aborigines and White Settlers in the Nineteenth Century. In C. T. Stannage (Ed.), *A New History of Western Australia*, pp. 72-123. Perth: University of Western Australia Press.

Gridneff, I. (2010). "PNG Miners' Strike at Ok Tedi Over Pay." April 8, Sydney Morning Herald. Retrieved from: http://news.smh.com.au/breaking-news-world/png-miners-strike-at-ok-tedi-over-pay-20100408-rtf0.html

GRI Global Reporting Initiative. (2010). Mining and Metals Sector Supplement. Retrieved from: http://www.icmm.com/page/36353/gri-mining-and-metals-sector-supplement

Hajkowicz S. A., Cook H, and Littleboy A. (2012). *Our Future World: Global Megatrends that Will Change the Way We Live.* The 2012 Revision. Canberra: CSIRO, Australia.

Hamann, R. and Kapelus, P. (2004). Corporate Social Responsibility in Mining in Southern Africa: Fair Accountability or Just Greenwash? *Development*, (47): 85–92. DOI: 10.1057/palgrave.development.1100056.

Harvey, B. and Brereton, D. (2005, 15 August). Emerging Models of Community Engagement in the Australian Minerals Industry. Paper presented to the UN conference on "Engaging communities" in Brisbane, Australia. Retrieved from http://www.riotinto.com/documents/Media-Speeches/UN_Conference_on_Community_Engagement_BH_150805.pdf

Harvey, D. (2008).Preface. In N. Smith, *Uneven Development: Nature, Capital, and the Production of Space.* Atlanta: University of Georgia Press.

Haslam-McKenzie, F. (2009). Farms and Mines: A Conflicting or Complimentary Land Use Dilemma in Western Australia? *J for Geog* 4(2), 113-128.

Hearn, G. J. (1995). Landslide and erosion mapping at Ok Tedi copper mine, Papua New Guinea. *Quarterly Journal of Mining Geology*, 28: 47-60.

Hess, M. (1994). Black and Red: The Pilbara Pastoral Worker's Strike 1946. *Aboriginal History*, 18(1): 65-83.

Hettler J., Irion G., and Lehmann, B. (1997). Environmental Impact of Mining Wast Disposal on a Tropical Lowland River System: A Case Study of the Ok Tedi Mine, Papua New Guinea. *Mineralium Deposita*, 32: 280-291. DOI: 10.1007/s001260050093.

Hill, R. (1995). Blackfellas and Whitefellas: Aboriginal Land Rights, the Mabo Decision, and the Meaning of Land. *Human Rights Quarterly*, 17(2):303. DOI: 10.1353/hrq.1995.0017.

Hilson, G. (2002). An Overview of Land Use Conflict in Mining Communities. *Land Use Policy*, 19: 65-73. DOI: 10.1016/S0264-8377(01)00043-6.

Hilson, G. (2006). Mining and Civil Conflict: Revisiting Grievance at Bougainville. *Minerals and Energy*, 2: 23-35. DOI: 10.1080/14041040601047937.

Holcombe, S. (2004). "Early Indigenous Engagement with Mining in the Pilbara: Lessons from a Historical Perspective." CAEPR (Centre for Aboriginal Economic Policy and Research) Working Paper No. 24. Canberra: ANU e-Press.

Holcombe, S. (2005). Indigenous Organizations and Mining in the Pilbara, Western Australia: Lessons from a Historical Perspective. *Aboriginal History*, 29: 107-135.

Holcombe, S. (2006). Community Development Packages: Development's Encounter with Pluralism in the Mining Industry. In T. Lea, E. Kowal and G. Cowlishaw (ed.), *Moving Anthropology: Critical Indigenous Studies*, pp. 79-94. Darwin: Charles Darwin University Press.

Holcombe, S. (2010). Sustainable Aboriginal Livelihoods and the Pilbara Mining Boom. In Ian Keen (ed.), *Indigenous Participation in Australian Economies: Historical and Anthropological Perspectives*, pp. 141-164. Canberra: ANU ePress.

Hoover, H. (1950). *Agricola: De Re Metallica*. New York: Dover Publications.

House of Representatives, Standing Committee on Regional Australia (2013). *Cancer of the Bush or Salvation of Our Cities? Fly-In, fly-Out and Drive-In, Drive-Iut Workforce Practices in Regional Australia*. Canberra: Commonwealth of Australia.

Howitt, R. (1997). "Lands, Rights, Laws: Issues of Native Title. "AIATSIS Native Title Research Unit, Regional Agreements Paper no 3. Canberra: AITSIS.

Howitt R. (1999). "Indigenous rights and regional economies: rethinking the building blocks." Paper presented to Rethinking Economy: Alternative Accounts Conference, August, Australian National University, Canberra.

Hughes, J., Emerson, S., and Hawes, R. (1997). Premiers Push Partial Extinguishment. West Australian, retrieved from: http://nfsa.gov.au/digitallearning/mabo/info/premiersPush-Partial.htm.

Hyndman, D. (1994). A Sacred Mountain of Gold: the Creation of a Mining Resource Frontier in Papua New Guinea. *J Pacific Hist*, 29(2): 203-221. DOI: 10.1080/00223349408572772.

Hyndman, D. (2001). Academic Responsibilities and Representation of the Ok Tedi Crisis in Postcolonial Papua New Guinea. *Contemporary Pacific*, 13(1): 35-54. DOI: 10.1353/cp.2001.0014.

HRW Human Rights Watch. (2011). Gold's Costly Dividend: Human Rights Impacts of Papua New Guinea's Porgera Gold Mine. Retrieved from: http://www.hrw.org/node/95776

ICMM. (2013). International Council of Mining and Minerals. Retrieved from: http://www.icmm.com/

ICMM International Council of Mining and Minerals. (2013a). Our History. Retrieved from: http://www.icmm.com/about-us/our-history

ICMM International Council of Mining and Minerals. (2013b). Community Development Toolkit (updated). Available from: http://www.icmm.com/community-development-toolkit

ICMM International Council of Mining and Minerals. (2013c). Sustainable Development Framework. Retrieved from: http://www.icmm.com/our-work/sustainable-development-framework

ICMM International Council of Mining and Minerals. (2013d). Member Companies. Retrieved from: http://www.icmm.com/members/member-companies

IIED International Institute for Environment and Development. (2002). Breaking New Ground: Mining, Minerals and Sustainable Development. Retrieved from: http://www.iied.org/mmsd-final-report

IISD International Institute for Sustainable Development. (2013). The Rise and Role of NGOs in Sustainable Development. Retrieved from: http://www.iisd.org/business/ngo/roles.aspx

Imhof, A. (1996). The Big Ugly Australian Goes to Ok Tedi. The Multinational Monitor, 17(3). Retrieved from: http://www.multinationalmonitor.org/hyper/mm0396.05.html

Inmet Mining Corporation. (2002). Inmet Announces Completion of Ok Tedi Arrangements. Retrieved from: http://ir.inmetmining.com/Files/1e/1edf8bd9-63f1-4507-990b-9918971aa13a.pdf

Independent State of PNG. (1995). Compensation (Prohibition of Foreign Legal Proceedings) Act 1995. Retrieved from: http://www.paclii.org/pg/legis/consol_act/coflpa1995499.rtf.

ISS Indigenous Support Services. (2001). "Agreements between Mining Companies and Indigenous Communities." A Report prepared for the Australian Minerals and Energy Environment Foundation. Retrieved from: INSERT

IUCN The World Conservation Union. (1991). *Caring for the Earth: A Strategy for Sustainable Living*. London: Earthscan.

Jackson, R. (1998). David and Goliath on the Fly (Book Review). *Journal of Pacific History*, 33(3): 307-311.

Jackson, R. (2003). "Muddying the Waters of the Fly: Underlying Issues or Stereotypes?" Working Paper 41, Research in Asia Pacific. Canberra: Research School of Pacific & Asian Studies, ANU.

Jackson, S. (2005). Indigenous Values and Water Resource Management: A Case Study from the Northern Territory, Australasian *J Env Mgmt* 12(3): 136-146. DOI: 10.1080/14486563. 2005.10648644.

Jackson, S. (2006). CompartmentaliZing Culture: The Articulation and Consideration of Indigenous Values in Water Resource Management, *Australian Geographer* 37(1): 19-32. DOI: 10.1080/00049180500511947.

Jenkins, R. (1987). *Transnationals and Uneven Development*. London: Croom Helm.

Jeschke, M. (2007). Mining Minerals & Sustainable Development. In *The A to Z of Corporate Social Responsibility*, Wiley, Hoboken, NJ, USA. Retrieved 11 July 2013 from: http://www. credoreference.com/entry/wileyazcsr/mining_and_minerals_for_sustainable_development_mmsd>

Johnson, S. L. and Wright, A. H. (2001). "Central Pilbara Groundwater Study. Water and Rivers Commission, Hydrogeological Record Series." Report HG 8. Perth: Water and Rivers Commission. Available online at: http://www.water.wa.gov.au/PublicationStore/first/12551_part1.pdf

Jordan, K. (2011). *The Australian Employment Covenant: A Research Update*. Centre for Aboriginal Economic Policy Research. Topical Issue No. 8/2011. Australian National University College of Arts and Social Sciences.

Jorgensen, D. (1981). Life on the Fringe: History and Society in Telefolmin. In R. Gordon (Ed.), *The Plight of Peripheral People in Papua New Guinea*, pp. 59-79. Cambridge, MA: Cultural Survival.

Jorgensen, D. (2004). Who and What is a Landowner? Mythology and Marking the Ground in a Papua New Guinea Mining Project. In A. Rumsey & J. Weiner (Eds), *Mining and*

Indigenous Lifeworlds in Australia and Papua New Guinea, pp. 68-100. Wantage, OXON: Sean Kingston Publishing.

Jorgensen, D. (2006). Hinterland History: The Ok Tedi Mine and its Cultural Consequences in Telefolmin. *Contemporary Pacific*, 18(2): 233-263. DOI: 10.1353/cp.2006.0021.

Kalinoe, N. (2008). "The Ok Tedi Mine Continuation Agreements: A Case Study Dealing with Customary Landowners' Compensation Claims." The National Research Institute Discussion Paper No. 105. Boroko, PNG: National Research Institute. Also available at: http://www.nri.org.pg/publications/Recent%20Publications/2010%20Publications/Discussion%20Paper%20105_OkTedi%5B1%5D.pdf

Kapelus, P. (2002). Mining, Corporate Social Responsibility and the "Community:" The Case of Rio Tinto, Richards Bay Minerals and the Mbonambi. *J Bus Ethics*, 39: 275-296. DOI: 10.1023/A:1016570929359.

Kay, P. (1995). PNG's Ok Tedi, Development and Environment. *Current Issues Brief*, No. 4. Canberra: Department of the Parliamentary Library.

Keane, B. (2010, June 15). Iron Ore's Raging Thirst Could Consume an Entire Industry. Crikey Media. Retrieved from: http://www.crikey.com.au/2010/06/15/iron-ores-raging-thirst-could-consume-an-entire-industry/

Kemp, D. (2009). Mining and Community Development: Problems and Possibilitiesof Local-Level Practice. *Comm Dev J* 45(2), 198-218. DOI: 10.1093/cdj/bsp006.

Kemp, D. and Keenan, J. (2009). Why Gender Matters. Rio Tinto. Retrieved from: http://www.riotinto.com/documents/ReportsPublications/Rio_Tinto_gender_guide.pdf

Kemp, D., Owen J. R., and Van de Graaff, S. (2012). Corporate Social Responsibility, Mining and the Audit Culture. *JCleaner Prod*, 24: 1-10. DOI: 10.1016/j.jclepro.2011.11.002.

Kirsch, S. (1989). Ok Tedi a Sewer. *Times of Papua New Guinea*, 1 June, p. 3.

Kirsch, S. (1993). The Yonggom of Papua New Guinea and the Ok Tedi Mine. In M. Miller (Ed.) *State of the Peoples: Global Rights Report on Societies in Danger*, p. 113. Boston: Beacon Press, 1993.

Kirsch, S. (1993b). "Social Impact of the Ok Tedi Mine on the Yonggom Villages of the North Fly, 1992." Ok-Fly Social Monitoring Project Report No. 5. Retrieved from: https://crawford.anu.edu.au/rmap/archive/Ok-Fly_social_monitoring/Ofsmp05-Kirsch1993-the-Yonggom-people.pdf

Kirsch, S. (1995). Social Impact of the Ok Tedi Mine on the Yonggom Villages of the North Fly, *Research in Melanesia* 19, pp. 23-102.

Kirsch, S. (1997). Is Ok Tedi a Precedent? Implications of the Lawsuit. In G. Banks & C. Ballard (Eds.), *The Ok Tedi Settlement: Issues, Outcomes and Omplications*, pp. 118-140. Canberra: National centre for Development Studies, Pacific Paper No. 27, RSPAS, Australian National University.

Kirsch, S. (2002). Anthropology and advocacy: a case study of the campaign against the Ok Tedi mine. *Critique of Anthropology*, 22(2): 175-200.

Kirsch, S. (2006). *Reverse Anthropology: Indigenous Analysis of Social and Environmental Relations in New Guinea*. Stanford, CA: Stanford University Press.

Kirsch, S. (2007). Indigenous Movements and the Risks of Counterglobalization: Tracking the Campaign against Papua New Guinea's Ok Tedi Mine. *Amer Ethnol*, 34(2): 303-321. DOI: 10.1525/ae.2007.34.2.303.

Kirsch, S. (2010). Sustainable Mining. *Dialectical Anthropol*. 34: 87-93. DOI: 10.1007/s10624-009-9113-x.

Knapp, A. (1998). *Social Approaches to an Industrial Past: The Archaeology and Anthropology of Mining*. London: Routledge. DOI: 10.4324/9780203265291.

Kranzberg, M. and Smith, C. S. (1979). Materials in History and Society (Part 1). *Mat Sci Eng*, 37(1): 1-39. DOI: 10.1016/0025-5416(79)90182-4.

Lahiri-Dutt, K. (Ed.). (2011a). *Gendering the Field: Toward Sustainable Livelihoods for Mining Communities*. Canberra: ANU E Press.

Lahiri-Dutt, K. (2011b). Digging Women: Toward a New Agenda for Feminist Critiques of Mining. *Gender, Place and Culture*, 1: 1-20.

LAMMP Latin American Mining Monitoring Program. (2013). About Us. Retrieved from: http://lammp.org/?page_id=17.

Lamphere, L. (2003). The Prospects and Perils of an Engaged Anthropology. *Soc Anthropol*, 11(2): 153-168. DOI: 10.1017/S0964028203000120.

Langton, M., Palmer, L, Tehan, M., and Shain, K. (Eds). (2004). *Honour among Nations? Treaties and Agreements with Indigenous People*. Melbourne: Melbourne University Press

Langton, M. (2008). The End of Big Men Politics. *Griffith Review* (22). Retrieved from: https://griffithreview.com/images/stories/edition_articles/ed22_pdfs/langton_ed22.pdf

Langton, M. (2010). The Resource Curse: New Outback Principalities and the Paradox of Plenty. *Griffith Review*, 28(Autumn): 47-83.

116 REFERENCES

Langton, M. (2012). Carving a Route to Indigenous Wealth. *Sydney Morning Herald* (November 17, 2012). Retrieved from: http://www.smh.com.au/national/carving-a-route-to-indige-nous-wealth-20121116-29hgj.html

Langton, M. and Longbottom, J. (Eds). (2012). *Community Futures, Legal Architecture: Foundations for Indigenous Peoples in the Global Mining Boom.* Abingdon, Oxon: Routledge.

Lankton, L. (1991). *Cradle to Grave: Life, Work, and Death at the Lake Superior Copper Mines.* New York: Oxford University Press.

Lannin, S. (2013). Gold Producer Newmont Mining is Cutting Jobs at its Boddington Mine in WA. *Australian Broadcasting Commission News*, Friday 26 July. Retrieved from: http://www.abc.net.au/news/2013-07-26/newmont-cuts-jobs-at-wa-mine/4847028

LMN (London Mining Network). (2013). About. Retrieved from: http://londonminingnetwork.org/about/.

Lourie, J. and Taufa, T. (1986). Ok Tedi Health and Nutrition Project, Papua New Guinea: Physique, growth and nutritonal status of the Wopkaimin of the Star Mountains. *Ann of Hum Biol*, 13(6): 517-536. DOI: 10.1080/03014468600008701.

Low, N. and Gleeson, B. (1998). Situating Justice in the Environment: The Case of BHP at the Ok Tedi Copper Mine. *Antipode*, 30(3): 201-226. DOI: 10.1111/1467-8330.00075.

Lowenhaupt Tsing, A. (2003). Natural Resources and Capitalist Frontiers. *Econ Pol Wkly*, 38 (48): pp. 5100-5106.

Lynch, M. (2002). *Mining in World History.* London: Reaktion Books.

MacCallum, J. Palmer, D. Wright, P., Cumming-Potvin, W., Brooker, M., and C. Tero. (2010). Australian Perspectives: Community Building through Intergenerational Exchange Programs. *J IntergenRelationships*, 8 (2): 113-127. DOI: 10.1080/15350771003741899.

MacCallum, J. Palmer, D., Wright, P., Cumming-Potvin, W., Northcote, J., Brooker, M., and Tero, C. (2006). "Community Building through Intergenerational Exchange Programs." Report to the National Youth Affairs Research Scheme (NYARS). Retrieved from: http://foi.deewr.gov.au/system/files/doc/other/community_building_through_intergen-erational_exchange_programs.pdf

Macdonald, I. and Rowland, C. (Eds). (2002). Tunnel Vision: Women, Mining and Communities. Melbourne: Oxfam Community Aid Abroad. Also available at: https://www.oxfam.org.au/wp-content/uploads/2011/11/oaus-tunnelvisionwomenmining-1102.pdf

Macintyre, M. (2001). Taking Care of Culture: Consultancy, Anthropology and Gender Issues. *Soc Analysis*, 45, 108-120.

Macintyre, M. and Foale, S. (2004). Politicized Ecology: Local Responses to Mining in Papua New Guinea. *Oceania*, 74(3): 231-251. DOI: 10.1890/03-0579.

MAC Mines and Communities. (2001). London Declaration. Retrieved from: http://www.mine-sandcommunities.org/article.php?a=8245

MAC Mines and Communities. (2013). Mines and Communities. Retrieved from: http://www.minesandcommunities.org/

McNair-Holland, L. (2006). *Breaking New Ground: Stories of Mining and Aboriginal People in the Pilbara*. Perth: Rio Tinto.

Measham, T., Haslem McKenzie, F., Moffat, K., and Franks, D. (2013). Reflections on the Role of the Resources Sector in Australian Economy and Society During the Recent Mining Boom. *Rural Soc J*, 22(2): 184-194. DOI: 10.5172/rsj.2013.22.2.184.

Mining Journal (2014). Mining 101. Retrieved from: http://www.mining-journal.com/knowledge/Mining-101

Mills, D. and Ratcliffe, R. (2012). After Method? Ethnography in the Knowledge Economy. *Qualitative Res*, 12(2): pp.147-164. DOI: 10.1177/1468794111420902.

Miskimin, H. A. (1977). *The Economy of Later Renaissance Europe 1460-1600*. Oxford: Cambridge University Press.

Morris, R. (2012). *Scoping Study: Impact of Fly in Fly out/Drive in Drive out Work Practices on Local Government*, Australian Centre of Excellence for Local Government, University of Technology, Sydney.

MMSD Mining, Minerals & Sustainable Development. (2002). *Ok Tedi Riverine Study (Appendix H)*. London: International Institute for Environment and Development (IIED).

MPI Minerals Policy Institute. (2013). Ok Tedi. Retrieved from: http://www.mpi.org.au/our-work/papua-new-guinea/ok-tedi/

Mudd, G. M. (2007). Global Trends in Gold Mining: Toward Quantifying Environmental and Resource Sustainability. *Resources Policy*, 32: 42-56. DOI: 10.1016/j.resourpol.2007.05.002.

Mudd, G. M. (2010). The Environmental Sustainability of Mining in Australia: key mega-trends and looming constraints. *Resources Policy* 35: 98-115. DOI: 10.1016/j.resourpol.2009.12.001.

Neate, B. (2013). Now and Then: Moola Bulla Station. ABC Open (audiovisual). Retrieved from: https://open.abc.net.au/posts/now-and-then-moola-bulla-50ga3yn#credits

Nelson, H. (2000). Liberation: The End of Australian Rule in Papua New Guinea. *J Pacific Hist*, 35(3): 269 -280. DOI: 10.1080/00223340020010562.

Newcrest. (2012a). Telfer Operations. Retrieved from: http://www.newcrest.com.au//operations. asp?category=6

Newmont. (2013). Aboriginal Agreement Forged. Retrieved from: http://www.newmont.com/features/our-communities-features/Aboriginal-Agreement-Forged

Ngarda (2012). Ngarda: Australia's Largest Aboriginal contractor. Retrieved from: http://www.ngarda.com.au/

Ngiragu, J. O. (1992). A History of Global Metal Pollution. *Science*, 272 (5259): 223-4.

O'Faircheallaigh, C. (1995). *Negotiations between Mining Companies and Aboriginal Communities: Process and Structure*. Centre for Aboriginal Economic Policy and Research. Report 86. Canberra: ANU Press. Retrieved from http://caepr.anu.edu.au/Publications/DP/1995DP86.php

O'Faircheallaigh, C. (2012). Women's Absence, Women's Power: Indigenous Women and Negotiations with Mining Companies in Australia and Canada. *Ethnic and Racial Studies*, 1: 1-19.

Ogan, E. (1991). The Cultural Background to the Bougainville Crisis. *J de la Société des Océanistes*, 92(1): 61-67. DOI: 10.3406/jso.1991.2897.

Olive, N. (2007). *Enough is Enough: A History of the Pilbara Mob*. Fremantle: Fremantle Arts Press.

Oliver, S. (2012). Is Australia's Resources Boom Over? Retrieved from: http://www.bbc.co.uk/news/business-19389211

OTML Ok Tedi Mining Consortium Limited. (2001). Key Statistics. Retrieved from archived reference: http://web.archive.org/web/20060820172746/http://www.oktedi.com/aboutus/keyStatistics.php

OTML Ok Tedi Mining Consortium Limited. (2008a). Diagram of operations. Retrieved from: http://www.oktedi.com/index.php?option=com_content&view=article&id=56&Itemid=65

OTML Ok Tedi Mining Consortium Limited. (2008b). Sustainability: Introduction. Retrieved from: http://www.oktedi.com/index.php?option=com_content&view=article&id=59&Itemid=68

Overlack, P. (1999). Bless the Queen and Curse the Colonial Office: Australasian Reaction to German Consolidation in the Pacific, 1879-1899. *J Pacific Hist*, 3(2): 133-152.

O'Connor, M. (Ed.). (1994). *Is Capitalism Sustainable: Political Economy and the Politics of Ecology*. New York: The Guildford Press.

Oxfam Australia (2010). The Role of Government. Retrieved from: https://www.oxfam.org.au/explore/mining/the-role-of-government/.

Oxfam (n.d.). Impacts of Mining. Retrieved from: https://www.oxfam.org.au/explore/mining/impacts-of-mining/

Papua New Guinea Mine Watch. (2011). Mining Suspension Enters Fourth Week at Ok Tedi. Retrieved from: http://ramumine.wordpress.com/2011/06/22/mining-suspension-enters-fourth-week-at-ok-tedi/

Papua New Guinea Mine Watch. (2013). About PNG Mine Watch. Retrieved from: http://ramumine.wordpress.com/about/

Pascoe, A. (2010). Skyscrapers gleam in Australia's "economic epicentre." Australian Federated Press. Available from: http://www.google.com/hostednews/afp/article/ALeqM5itsPwa76mGISoHXLlfRu5wMwg2-g

Perez, F. and Sanchez, L. E. (2009). Assessing the Evolution of Sustainability Reporting in the Mining Sector. *Env Mgmt* 43(6), 949-961. DOI: 10.1007/s00267-008-9269-1.

Pernetta, J. (Ed). (1988). "Potential Impact of Mining on the Fly River." United National Environment Program Regional Seas Report and Studies No 99. Nairobi: UNEP.

Pernetta, J. (1988). The Ok Tedi Mine: Environment, Development and Pollution Problems. In J. Pernetta (Ed). *Potential Impacts of Mining on the Fly River*, pp. 1-8. United National Environment Program Regional Seas Report and Studies No 99. Nairobi: UNEP.

Petrova, S. and Marinova, D. (2013). Social Impacts of Mining: Changes within the Local Social Landscape. *Rural Soc J.* 22(2), 153-165. DOI: 10.5172/rsj.2013.22.2.153.

PHIDU Public Health Information Development Unit. (2005). Population Health Profile of the Pilbara. Population Profile Series: No. 112. Adelaide: Public Health Information Development Unit (PHIDU). Retrieved from: http://www.publichealth.gov.au/pdf/profiles/2005/614_Pilbara_DGP.pdf

Pick, D., Dayaram, K., and Butler, B. (2008). Neo-Liberalism, Risk and Regional Development in Western Australia: The Case of the Pilbara. *Intl J Sociol and Soc Policy*, 28(11/12): 516-527. DOI: 10.1108/01443330810915224.

Pintz, W. (1984). *Ok Tedi: Evolution of a Third World Mining Project*. London: Mining Journal Books.

PNGSDP Papua New Guinea Sustainable Development Program. (2013). Company Profile. Retrieved from: http://www.pngsdp.com/index.php/company-profile

Polanyi, K. (1957). *The Great Transformation*. Boston: Beacon Press.

Postan, M. (1987). *The Cambridge Economic History of Europe: Trade and industry in the Middle Ages*. Cambridge: Cambridge University Press.

Public Health Assocation, Australia (2012). WA Branch FIFO Forum: Understanding the Impact of FIFO on the Well Being and Health http://www.phaa.net.au/documents/130130PHAA_FIFO%20Forum%20report_2012.pdf (accessed 28 June, 2013)

Radetzki, M. (2009). Seven Thousand Years in the Service of Humanity – The History of Copper, the Red Metal. *Resources Policy*, 34(4): 176-184. DOI: 10.1016/j.resourpol.2009.03.003.

Rajak, D. (2010). From Boardrooms to Mineshafts: In Pursuit of Global Corporate Citizenship. In P. Von-Hellermann and S. Coleman (eds), *Problems and Challenges in Multi-sited Ethnography*. London: Routledge.

Rankin, W. (2011). *Minerals, metals and sustainability meeting future material needs*. Collingwood, VIC: CSIRO.

Reid, D. (1995). *Sustainable Development: An Introductory Guide*. London: Earthscan.

Regan, A. J. (1998). Causes and Course of the Bougainville Conflict. *J Pacific Hist* 33(3), 269-285. DOI: 10.1080/00223349808572878.

Riley, D. (2008). *Engineering and Social Justice*. San Rafael, CA: Morgan & Claypool Press. DOI: 10.2200/S00117ED1V01Y200805ETS007.

Riley, D. (2012). We've Been Framed! Ends, Means, and the Ethics of the Grand(iose) Challenges. *Intl J Eng, Soc Justice and Peace*, 1(2): 123-136.

Rio Tinto. (2007). Our contribution to Western Australia; community investment review. Retrieved from: http://www.riotinto.com/documents/ReportsPublications/rtwa_community_investment_review_.pdf

Ritter, D. (2002). The Fulcrum of Noonkanbah. *J Australian Studies*, 26: 51-58. DOI: 10.1080/14443050209387803.

Robins, N. and Hagan, N. (2012). Mercury Production and Use in Colonial Andean Silver Production: Emissions and Health Implications. *Env Health Perspectives*, 120(5): 627-631. DOI: 10.1289/ehp.1104192.

Robbins, P. (2011). *Political Ecology: A Critical Introduction*. London: John Wiley & Sons.

Ross, M. (2004). "Mineral Wealth & Equitable Development." Background papers, World Development Report No. 6 Equity & Development. Retrieved from: https://openknowledge.worldbank.org/bitstream/handle/10986/9173/WDR2006_0017.pdf?sequence=1

Rostovtzeff, M. (1957). *The Social and Economic History of the Roman Empire*. Oxford: Clarendon Press.

Rudenno, V. (2004). *Mining Valuation Handbook*. Camberwell, VIC: Wrightbooks.

Rumley, H. and Barber, K. (2004). "We Used to Get Our Water Free." A Study and Report prepared for the Water and Rivers Commission of Western Australia, Contract 7-57044-3. Perth: Water and Rivers Commission. Also available at: www.water.wa.gov.au/PublicationStore/first/80735.pdf

Sachs, W. (1999). Sustainable Development and the Crisis of Nature: On the Political Anatomy of an Oxymoron. In F. Fischer and M. A. Hajer (Eds), *Living With Nature: Environmental Politics as Cultural Discourse*, pp.23-41. Oxford: Oxford University Press. DOI: 10.1093/019829509X.003.0002.

Schuurkamp, G. S. (1990). Diethylcarbamazine in the control of bancroftian filariasis in the highly endemic Ok Tedi area of Papua New Guinea: phase 1. *PNG Medical Journal*, 33(2):89-98. DOI: 10.1016/0035-9203(92)90097-V.

Schwab, R. (2006). "Discussion Paper 100/1995 Centre of Aboriginal Economic Policy Research." Australian National University. Canberra, ACT.

Sethi, S. P. and Emelianova, O. (2006). A Failed Strategy of Using Voluntary Codes of Conduct by the Global Mining Industry. *Corp Gov*, 6(3): 226-238. DOI: 10.1108/14720700610671837.

Shire of Boddington. (2004). History. Retrieved from: http://www.boddington.wa.gov.au/history

Slater, M. (2013). Hippies at the Gate: Why the Left and Right Hate Coal Seam Gas. Retrieved from: http://www.crikey.com.au/2013/02/25/hippies-at-the-gate-why-the-left-and-right-hate-coal-seam-gas/?wpmp_switcher=mobile

SMH Sydney Morning Herald. (2012). Pain and Privilege in a Boom and Gloom Economy. February 12, 2012. Retrieved from: http://www.smh.com.au/federal-politics/editorial/pain-and-privilege-in-a-boom-and-gloom-economy-20120217-1teeg.html?skin=text-only

SMIT Star Mountains Institute of Technology. (2013). Tabubil. Retrieved from: http://www.smit.ac.pg/tabubil-town

Smith, A. B. (2002). *Under a Bilari Tree I Born*. Fremantle Arts Centre Press Fremantle, Western Australia.

Stiglitz, J. (2006). *Making Globalization Work*. New York: W.W. Norton & Co.

Spitz, K. and Trudinger, J. (2009). *Mining and the Environment: From Ore to Metal*. London: Taylor & Francis.

Standing Committee on Regional Australia. (2013). "Inquiry into the use of 'fly-in, fly-out' (FIFO) workforce practices in regional Australia. Parliament of Australia." Retrieved from: http://www.aph.gov.au/parliamentary_business/committees/house_of_representatives_committees?url=ra/fifodido/report.htm

Stanfield, J. R. (1989). Karl Polanyi and Contemporary Economic Thought. *Rev Soc Econ*, 47(3): 266-279. DOI: 10.1080/00346768900000027.

Stocking, G. (1992). Paradigmatic Traditions in the History of Anthropology. In G. Stocking (Ed.) *The Ethnographer's Magic and Other Essays in the History of Anthropology*, pp. 342-361. Madiscon, WI: University of Wisconsin Press.

Storey, K. (2010). Fly-in/Fly-out: Implications for Community Sustainability. *Sustainability*. 2:1161-1181. DOI: 10.3390/su2051161.

Su, R. (2013). Newcrest Mining Cuts Jobs at Telfer, More Mines to Have Same Fate. *International Business Times*, July 11 2013. Retrieved from: http://au.ibtimes.com/articles/488981/20130711/newxrest-mining-australia.htm

Taukuro, B.D. (1980). The WHO North Fly Clinico-Epidemiological Pilot Study. *PNG Med J*, 23: 80-86.

Taylor, J. and Scambary, B. (2005). "Indigenous People and the Pilbara Mining Boom: A Baseline for Regional Participation." Centre for Aboriginal Economic Policy Research (CAEPR), Research Monograph No. 24. Canberra: ANU E-Press.

Taylor, J. and Simmonds, J. (2009). Family stress and coping in the fly-in fly-out workforce. *Australian Comm Psychol*, 21(2): 23-36.

The Watut Cries. (2012). The Watut Cries Blogspot. Retrieved from: http://thewatutcries.blogspot.com.au/.

Thomason, J. and Hancock, M. (2011). "PNG Mineral Boom: Harnessing the Extractive Sector to Deliver Better Health Outcomes." Development Policy Centre Discussion Paper #2, Crawford School of Public Policy, The Australian National University, Canberra.

Tonkinson, R. (1991). *The Mardu Aborigines: Living the Dream in Australia's Western Desert.* Forth Worth TX: Holt, Rhinehart & Winston.

Tonkinson, R. (2007a). Aboriginal "Difference" and "Autonomy" Then and Now: Four Decades of Change in a Western Desert Community. *Anthropological Forum*, 17(1): 41-60. DOI: 10.1080/00664670601168476.

Tonkinson, R. (2007b). From Dust to Ashes: The Challenges of Difference. *Ethos*, 72(4): 509-534.

Trenwith, C. (2012, October 2). Perth's Rents Smash Major Capital Cities. Retrived from *WA Today*: http://www.watoday.com.au/wa-news/perth-rents-smash-major-capital-cities-20121010-27dtk.html

Tonts, M., Plummer, P., and Lawrie, M. (2013). Socio-Economic Wellbeing in Australian Mining Towns: A Comparative Analysis. *J Rural Studies*, 28: 288-301. DOI: 10.1016/j.jrur-stud.2011.10.006.

Townsend, W. (1988). Giving Away the Fly River. In J. Pernetta (Ed.), *Potential Ompact of Mining on the Fly River*, pp. 107-119. United National Environment Program Regional Seas Report and Studies No 99. Nairobi: UNEP.

Townsend, P. K. and Townsend, W. H. (2004). Assessing an Assessment: The Ok Tedi Mine. Bridging Scales and Epistemologies: Linking Local Knowledge and Global Science in Multi-Scale Assessments, Conference proceeding, Alexandria, Egypt, March 17-20. Also available at: http://www.unep.org/maweb/documents/bridging/papers/townsend.patricia.pdf.

Tylecote, R. (1992). *A History of Metallurgy* (2nd Edition). London: Institute of Minerals. First published 1976.

Ulijaszek, S. J., Hyndman D. C., Lourie J. & Pumuye, A. (1987). Mining, modernisation and dietary change among the Wopkaimin of Papua New Guinea. *Ecology of Food and Nutrition*, 20(2): 143-156.

UNEP. (1992). Rio Declaration on Environment & Development. Retrieved from: http://www.unep.org/Documents.Multilingual/Default.asp?documentid=78&articleid=1163

UNEP (2013). Waste from Consumption and Production—The Ok Tedi Case: A Pot of Gold. In *Vital Waste Graphics*, a joint publication between UNEP & Grid-Arendal. Retrieved from: http://www.grida.no/publications/vg/waste/page/2859.aspx

UNESCO. (2010). *Engineering: Issues, Challenges and Opportunities for Development*. Paris: Unesco Publishing.

United Nations. (1987). Report of the World Commission on Environment and Development: Our Common Future. Retrieved from: http://conspect.nl/pdf/Our_Common_Future-Brundtland_Report_1987.pdf

Urkidi, L. (2010). A Global Environmental Movement against Gold Mining: Pascua-Lama in Chile. *Ecological Economics*, 70(2): 219-227. DOI: 10.1016/j.ecolecon.2010.05.004.

Urkidi, L. (2011). The Defence of Community in the Anti-Mining Movement of Guatemala. *J Agrarian Change*, 11(6): 556-580. DOI: 10.1111/j.1471-0366.2011.00326.x.

Vela, B. (1975). *Tales of Potosi*. Providence University Press.

Visvanathan, S. (1997). Mrs. Brundtland's Disenchanted Cosmos. In G. Tuathail & S.Dalby (Eds), *Geopolitics Reader*. London: Routledge.

Walker, P. (2006). Political Ecology: Where is the Policy? *Progress in Human Geography*, 30(3): 382-395. Also available on: http://politikon.decyvpoliticas-unam.org/respaldo/Desktop/carpetas%20destok/Documents/descargas/PoliticalEcology_PWalker.pdf

Walker, W. (2009). Pilbara. In J. Gregory & J. Gothard (Eds), *Historical Encyclopedia of Western Australia*, pp. 691-693. Perth: University of Western Australia Press.

Warhurst, A. (2002). *Sustainability Indicators and Sustainability Performance Management*. London: Mining & Minerals for Sustainable Development (MMSD). Also available at: http://www.commdev.org/userfiles/files/681_file_sustainability_indicators.pdf

WCED (World Commission on Environment & Development). (1987). Our Common Future. Retrieved from: http://conspect.nl/pdf/Our_Common_Future-Brundtland_Report_1987.pdf

Webster-Smith, B. (1965). 60 Centuries of Copper. Retrieved from: http://www.coppergroupint.com/admin/descargas/16e.pdf

Whitmore, A. (2005). The Emperor's New Clothes: Mining and Sustainable Development. In B. Marker (Ed.), *Sustainable Minerals Operations in the Developing World*, pp. 233-242. London: Geological Society.

Williams, R. (2012, December 3). How the Resources Boom Split the Golden State. *Sydney Morning Herald*. Retrieved from: http://www.smh.com.au/business/how-the-resources-boom-split-the-golden-state-20111202-1obaw.html

Wilson, J. (1980). The Pilbara Aboriginal Social Movement: An Outline of its Background and Significance. In R. Berndt and C. Berndt (Eds), *Aborigines of the West: Their Past and their Present*, pp. 151-166. Perth: University of Western Australia Press.

Worsley Alumina. (1999). *Worsley Alumina Boddington Gold Mine project : flora and fauna studies*. Boddington W.A.: Worsley Alumina.

WWF (World Wildlife Fund). (2002). Living Plant Report 2002. Retrieved from: http://wwf.panda.org/about_our_earth/all_publications/living_planet_report/living_planet_report_timeline/lpr02/.

YAC, Yindjibarndi Aboriginal Corporation. (2011). The Yingjibarndi Story. Retrieved from: http://www.youtube.com/watch?v=VGZ_H_SWiQM

Author Biographies

Rita Armstrong is an anthropologist with a Ph.D. from the University of Sydney, based on two year's fieldwork in a longhouse community in Central Borneo. With an undergraduate major in History, she combines historical research with anthropological methodologies and interests to analyze a variety of issues: Indigenous perceptions of social change, political economy of the interaction between shifting cultivators and the state, subjective understanding of "development" and how all these influence and shape local identity. She has worked with Caroline Baillie, an engineer and social activist, for a number of years in developing interdisciplinary teaching material for first-year engineers at the University of Western Australia, and, most recently, on research projects funded by the International Mining for Development Centre. She continues to teach in Anthropology and Engineering and this experience has underlined the importance of developing collaborative research projects across these disciplines to better understand how we can resolve the increasing inequity in peoples' capacity to deal with issues such as climate change, resource extraction, and diminishing water supply.

Caroline Baillie is Chair of Engineering Education for the Faculty of Engineering, Computing and Mathematics at the University of Western Australia. Before coming to Perth, Caroline was Chair of Engineering Education Research and Development at Queens University, Kingston, Ontario, and she also held posts at Imperial College and the University of Sydney.

Caroline is particularly interested in ways in which science and engineering can help to co-create solutions for the environment as well as social problems. She founded the global "Engineering and Social Justice" network (esjp.org) in 2004 and applies this lens to her own technical work on low cost natural fiber composites for developing countries. Her not-for-profit organization "Waste for Life" (wasteforlife.org) works to create poverty-reducing solutions to environmental issues. Caroline is Editor of this series "Engineers, Technology and Society."

Dr. Wendy Cumming-Potvin is a fulltime academic at Murdoch University's School of Education in Western Australia. Born and raised in Canada, Wendy has lived, worked, and studied in Australian mining communities. Her expertise lies in qualitative research promoting social justice in literacies, communities of practice, and engineering. As a project leader in a Cooperative Research Centre project, Wendy is examining the use of technology for promoting socially inclusive communities. She was recently awarded a Vice Chancellor's Award for Excellence in Teaching at MU.9